MIOMBO ECOLOGY AND MANAGEMENT

MIOMBO ECOLOGY AND MANAGEMENT

An Introduction

EMMANUEL N. CHIDUMAYO

IT PUBLICATIONS *in association with the STOCKHOLM ENVIRONMENT INSTITUTE, 1997*

Practical Action Publishing Ltd
25 Albert Street, Rugby, CV21 2SD, Warwickshire, UK
www.practicalactionpublishing.com

in association with
The Stockholm Environment Institute

©Stockholm Environment Institute, 1994, 1997

First published by Intermediate Technology Publications Ltd in 1997
Transferred to digital printing 2008

ISBN 10 Paperback: 1853394114
ISBN 13 Paperback: 9781853394119
ISBN Library Ebook: 9781780445496
Book DOI: http://dx.doi.org/10.3362/9781780445496

All rights reserved. No part of this publication may be reprinted or reproduced or utilized in any form or by any electronic, mechanical, or other means, now known or hereafter invented, including photocopying and recording, or in any information storage or retrieval system, without the written permission of the publishers.

A catalogue record for this book is available from the British Library.

The authors, contributors and/or editors have asserted their rights under the Copyright Designs and Patents Act 1988 to be identified as authors of their respective contributions.

Since 1974, Practical Action Publishing has published and disseminated books and information in support of international development work throughout the world. Practical Action Publishing is a trading name of Practical Action Publishing Ltd (Company Reg. No. 1159018), the wholly owned publishing company of Practical Action. Practical Action Publishing trades only in support of its parent charity objectives and any profits are covenanted back to Practical Action (Charity Reg. No. 247257, Group VAT Registration No. 880 9924 76).

Typeset by Dorwyn Ltd, Rowlands Castel, Hants, UK

Contents

LIST OF TABLES	vii
LIST OF FIGURES	x
LIST OF TEXT BOXES	xiii
LIST OF EQUATIONS	xiv
FOREWORD	xv
ACKNOWLEDGEMENTS	xvii
ABBREVIATIONS	xviii

1 INTRODUCTION — 1

THE ZAMBEZIAN PHYTOREGION — 1
- Dry forest — 1
- Swamp and riparian forest — 3
- Woodland — 3
- Thicket — 4
- Grassland — 4

THE ECOLOGY OF MIOMBO WOODLAND — 6
- Climate — 6
- Soils — 9
- Nutrient cycling — 14
- Soil conservation — 16
- Water use and conservation — 18
- Tree regeneration — 19
- Woody plant biomass production — 30

2 THE UTILIZATION OF MIOMBO — 35

AGRICULTURE — 36
- Cultivation — 36
- Pastoralism — 41

WOOD PRODUCTS — 43
WOODFUEL — 46
WILD FOODS — 50
MEDICINAL PLANTS — 53
ANIMAL PRODUCTS — 55

Bush meat	54
Edible insects	55
Honey	57

3 THE ROLE OF FIRE — 58

THE OCCURRENCE OF FIRE — 58
CAUSES AND FREQUENCY OF FIRES — 62
PLANT ADAPTATIONS TO FIRE — 64
THE EFFECTS ON SUCCESSION — 69

4 MANAGEMENT INVENTORIES AND ASSESSMENTS — 73

INVENTORIES — 73

- Objectives and procedure — 73
- Sampling design and plots — 77
- Types of data — 83
- Site history — 87

ASSESSMENTS — 88

- Plant biomass — 88
- Species diversity — 98
- Seeds and seedlings — 102
- Cultivation and deforestation — 107
- Critical population density under shifting cultivation — 108
- Charcoal yield and deforestation — 111
- The effects of fire — 116
- The effects of wood carbonization on soil — 120
- The effects of deforestation — 120
- Grazing and browsing — 122
- Litter decomposition — 124

5 MANAGEMENT GUIDELINES — 127

NATURAL REGENERATION — 128

- Harvesting techniques — 128
- Cutting cycles — 131
- Fire management — 132
- Traditional tree conservation practices — 133
- Grazing and browsing — 134

PLANTATION REGENERATION — 136

- Species selection — 136
- Seed collection and storage — 139
- Seed sowing — 140
- Nursery techniques — 141
- Vegetative propagation — 143

	Planting and tending	143
	Enrichment planting on charcoal spots	143
	CATCHMENT AND BIODIVERSITY CONSERVATION	144
References		147
Appendices		157

List of tables

Table 1.1	Distribution of land cover types in Zambia by province	6
Table 1.2	Climatic seasons at Lusaka, central Zambia	8
Table 1.3	Mean species diversity in miombo subtypes in Zambia	9
Table 1.4	Average textural composition of dry miombo soils in central Zambia	9
Table 1.5	Edaphic preferences by *Brachystegia* species in the Zambian Copperbelt	13
Table 1.6	Soil organic matter content in dry and wet miombo	13
Table 1.7	Average concentration of exchangeable bases in dry and wet miombo top soil	13
Table 1.8	Woody plant species diversity in 0.10 ha sample plots in some Zambian forests	14
Table 1.9	Differential nutrient concentration in plant tissues of different plants in miombo woodland	15
Table 1.10	Nitrogen and phosphorus concentration in foliage and leaf litter of miombo trees	15
Table 1.11	Changes in nutrient stocks in tree leaf and grass litter in dry miombo during the rainy season	17
Table 1.12	Diversity of fruit and seed dispersal types among miombo woodland trees and shrubs	21
Table 1.13	Seedling emergence rates from seeds stored for varying periods under uncontrolled room conditions in five miombo woodland species	23
Table 1.14	Shoot growth during the establishment period of miombo woodland tree seedlings	25
Table 1.15	Changes in sucker shoots per plant during coppice establishment in dry miombo	27
Table 1.16	Woody plant density one year after cutting dry miombo	28
Table 1.17	Stem height growth in regrowth dry miombo	28
Table 1.18	Above ground woody plant biomass in some Zambezian vegetation types in Zambia	29
Table 2.1	Distribution of human population in Zambia in 1990 by residence	35

Table 2.2	Above ground woody plant biomass portioning for chitemene cultivation in northern Zambia	37
Table 2.3	Trends in woodland conversion to chitemene in a 250-ha block of miombo in Bwacha village in Kasama district, Northern Province, Zambia	37
Table 2.4	Changes in top-soil nutrient pool during the fundikila grass-mound cultivation cycle in north-eastern Zambia	38
Table 2.5	Cultivated areas by vegetation and traditional farming system in Zambia in 1990	41
Table 2.6	Average exploitable timber stocking rate in north-western wet miombo in the Zambian Copperbelt	42
Table 2.7	Tree species preferred for building and other purposes in miombo woodland by the Bemba of northern Zambia	45
Table 2.8	Average length and density of poles in central dry miombo in Zambia	45
Table 2.9	National energy supply and consumption in Zambia in 1990	47
Table 2.10	Household woodfuel consumption in Zambia in 1990 by residence	47
Table 2.11	Cord wood biomass utilization for charcoal production in central dry miombo in Zambia	49
Table 2.12	Temporal changes in the rate of deforestation in the Zambian Copperbelt	50
Table 2.13	Diversity and abundance of wild fruit trees in Zambian miombo	50
Table 2.14	Reputed miombo woodland medicinal trees	52
Table 2.15	The structure of woody plant biomass in old-growth dry miombo in central Zambia	53
Table 3.1	Above ground flammable biomass in miombo woodland during the dry season	57
Table 3.2	Some characteristics of wild fires in miombo woodland during August and September	62
Table 3.3	Perceptions about causes of wild fires among 1410 rural households in the miombo region of Zambia	62
Table 3.4	Leaf flush and dry season wood moisture content in common dry miombo woodland trees in central Zambia	65
Table 3.5	Effect of fire on trees at a 1988 selectively cut dry miombo site	68
Table 3.6	Average bark thickness of some dry miombo trees in central Zambia	68
Table 3.7	Trapnell's fire tolerance classification of trees and shrubs after 11 continuous years of fire treatments in wet miombo at Ndola in the Zambian Copperbelt	70
Table 3.8	Effects of long term (1933–82) fire exclusion and burning on species diversity in wet miombo in Zambia	71
Table 3.9	Changes in tree density after 11 years of fire treatments in wet miombo at Ndola in the Zambian Copperbelt	72

Table 4.1	Wood biomass estimates by forest type in Zambia	77
Table 4.2	Generalized biomass equations for estimating above ground woody plant biomass of small and large stem in miombo woodland	93
Table 4.3	Estimates of annual tree leaf production in central Zambia dry miombo during 1991–93	95
Table 4.4	Specific wood density of some miombo woodland trees	96
Table 4.5	Regression for estimating cord wood volume and biomass from basal area at breast height and biomass from stacked volume for different miombo woodland biomass classes	96
Table 4.6	Changes in moisture content of above ground green grass biomass in dry miombo woodland	97
Table 4.7	Chemical composition of plant biomass in woodland in central Zambia	98
Table 4.8	Two-by-two table for calculating species diversity	100
Table 4.9	Percentage similarity for three miombo samples in Central Zambia	102
Table 4.10	Annual variations in seed production in two tree species at four central Zambian dry miombo sites	104
Table 4.11	Seedling emergence rate from seeds of miombo trees sown in miombo woodland soil immediately after seed ripening or dispersal	105
Table 4.12	Trend in deforestation caused by cultivation in Zambia during 1969–90 and projections for the years 2000 and 2010	106
Table 4.13	Land categories and use in the chitemene cultivation region in Zambia	108
Table 4.14	Changes in households and average household size in Zambia	112
Table 4.15	Charcoal consumption and wood used in charcoal production in Zambia	114
Table 4.16	Estimated charcoal production by the earth-kiln method in different vegetation types in Zambia	115
Table 4.17	Assessments of early and late burning on stem mortality after two years of treatment at Mutupa plots in the Zambian Copperbelt	115
Table 4.18	Response of miombo species to clear cutting and fire management	119
Table 4.19	Changes in species diversity in wet miombo woodland under fire management at Ndola, Zambia	119
Table 4.20	The immediate effect of an August wild fire on a dry miombo soil at Chisamba in central Zambia	120
Table 4.21	Changes in the concentration of extractable phosphorus in charcoal kiln soil in dry miombo in central Zambia	121

Table 4.22	Average soil moisture content in uncut plots and plots clear-cut by stumping in 1990 at an old-growth dry miombo site in central Zambia	122
Table 4.23	The effect of mammalian herbivores on a wetland grassland on the Kafue Flats, Zambia	124
Table 5.1	Stem size of some miombo trees at the age of 10 years in coppiced regrowth and plantation plots in the Zambian Copperbelt	127
Table 5.2	Species diversity in uneven age old growth and coppiced regrowth miombo in Zambia	129
Table 5.3	Structure of reserved trees in cultivated plots in three miombo woodlands areas in Zambia	131
Table 5.4	Effect of timing of tree cutting on production during the first growing season	131
Table 5.5	Suggested cutting cycles for natural regeneration in miombo	132
Table 5.6	Mortality among marked trees at Ndola old-growth miombo plots maintained under different fire treatments for 11 years	132
Table 5.7	Some performance criteria to guide the selection of miombo trees for plantation regeneration	138
Table 5.8	Vegetation conservation in forest reserves and national parks in Zambia	138
Table 5.9	Changes in the size of area under forest reserves in Zambia during 1978–87	146

List of figures

Note: In most of the diagrams, data points are connected by lines. This is to identify more easily the different series, and does not imply continuous data points.

Figure 1.1	The Zambezian phytoregion of central and southern Africa	2
Figure 1.2	Central and southern African countries with miombo vegetation	2
Figure 1.3	Generalized vegetation map of Zambia	5
Figure 1.4	Climate diagrams for three stations in miombo region	7
Figure 1.5	Vegetation catena in miombo from dambo to interfluve crest	10
Figure 1.6a	Spatial successions of dominant species along a 160 m hillside transect in central dry miombo	11
Figure 1.6b	Spatial successions of dominant species along a 100 m flat interfluve crest transect in central dry miombo	11

Figure 1.7	Seasonality in rainfall and top soil moisture content in dry miombo	12
Figure 1.8	Monthly variation in ammonium and nitrate concentration in dry miombo top soil	12
Figure 1.9	Monthly variation in exchangeable sodium and potassium in dry miombo	14
Figure 1.10	Decomposition of wet miombo tree leaf litter	16
Figure 1.11	Crown projection in wet miombo	18
Figure 1.12	Monthly variation in dry miombo greenness	19
Figure 1.13	Components of river flow in Zambia	20
Figure 1.14	Flowering pattern in dry miombo	20
Figure 1.15	Annual frequency of fruiting in three miombo trees	21
Figure 1.16	Reproductive phenology of some dry miombo trees	22
Figure 1.17	Shoot growing season in seedlings of two miombo trees	24
Figure 1.18	Phosphorus content in *Afzelia quanzensis* seedlings grown in soil with differential concentration of available phosphorus	25
Figure 1.19	Early development of shoot and root in seedlings of *J. globiflora*	26
Figure 1.20	Early succession from uncut woodland to fourth year regrowth by dominant species after clearing a dry miombo plot in Zambia	29
Figure 1.21	Mean height growth of three miombo trees in regrowth dry miombo	30
Figure 1.22	Temporal changes in foliar nitrogen:phosphorus ratio in three miombo trees	31
Figure 1.23	Correlation between above ground wood biomass and age of coppiced dry miombo	32
Figure 1.24	Correlation between above ground wood biomass and age of coppiced wet miombo	33
Figure 1.25	Age-related changes in leaf production and mean wood annual increment in coppiced dry miombo	33
Figure 1.26	Age-related changes in leaf production and mean wood annual increment in coppiced wet miombo	34
Figure 2.1	Decline in maize yield and soil pH under monocropping with continuous application of nitrogenous fertilizer at Misamfu in northern Zambia	39
Figure 2.2	Monthly changes in nitrogen content in grass and tree leaves in dry miombo	40
Figure 2.3	Growth in cattle population during 1964–87 in Western Province, Zambia	42
Figure 2.4	Main commercial timber areas in Zambian indigenous forests	43
Figure 2.5	Teak (*Baikiaea plurijuga*) harvesting in south-western Zambia 1933–82	44

Figure 2.6	Timber harvesting in wet miombo in the Zambian Copperbelt 1947–60	44
Figure 2.7	Changes in stem size structure in dry miombo caused by selective felling	46
Figure 2.8	Seasonal availability of ripe edible wild fruits in Zambian miombo	51
Figure 2.9	Distribution of national parks and game management areas in Zambia	54
Figure 2.10	Decline in the elephant population in Zambia during 1960–93	55
Figure 2.11	Changes in marketed honey in North-western Province, Zambia, during 1975–91	56
Figure 3.1	Mean daily maximum temperature and relative humidity in miombo	58
Figure 3.2	Cumulative tree leaf litter fall in miombo	59
Figure 3.3	Seasonality in mean moisture content of standing dry grass and tree leaf litter in dry miombo	61
Figure 3.4	Effects of decomposition and fire on accumulation of tree leaf litter in dry miombo	61
Figure 3.5	Mean monthly frequency of 1203 forest plantation fires in the Zambian Copperbelt during 1975–93	63
Figure 3.6	Effect of August and September fires on leaf production in *Isoberlinia angolensis* and *Uapaca kirkiana*	67
Figure 4.1	Land use and vegetation cover types in the Zambian Copperbelt in 1984	75
Figure 4.2	Sample plot location at Mwekera	78
Figure 4.3	Effect of increasing sampling area on biomass estimate and standard deviation in dry miombo	80
Figure 4.4	Types of woodland inventory plots	81
Figure 4.5	Stem measurement positions and marking of stem	84
Figure 4.6	Textural triangle for determining soil texture classes	86
Figure 4.7	Root structure in *Brachystegia spiciformis*	92
Figure 4.8	Types of soil pits for assessing root biomass in miombo	94
Figure 4.9	Grass production in dry miombo	97
Figure 4.10	Species-area curves for three forest types in central Zambia	100
Figure 4.11	Pattern of seed dispersal by an isolated *Julbernardia globiflora* tree	104
Figure 4.12	Seedling emergence in three miombo trees	106
Figure 4.13	Initial survivorship of seedlings of *Isoberlinia angolensis* and *Uapaca kirkiana* seedlings at a dry miombo site	107
Figure 4.14	Correlation between population density and deforestation caused by chitemene shifting cultivation in Mansa district, Luapula Province, Zambia	109

Figure 4.15	Correlation between household size and annual charcoal consumption per household and per capita in urban Zambia	113
Figure 4.16	Increase in charcoal consumption in Zambia	113
Figure 4.17	Spatial and temporal patterns of deforestation in the Copperbelt	118
Figure 4.18	Mean stream flow and duration before (1969–73) and after clearing wet miombo (1975–79)	121
Figure 4.19	A seven-point scale for assessing bark removal below 2 m	123
Figure 4.20	Changes in fungal biomass in top soil in wet miombo at Misamfu in northern Zambia	125
Figure 5.1	Clear-cutting with shelterbelts in miombo woodlands	130
Figure 5.2	The effect of fire control on above ground wood biomass accumulation in regrowth dry miombo	134
Figure 5.3	The mortality of coppiced tree stumps caused by overbrowsing by cattle in experimental wet miombo paddocks	136
Figure 5.4	Mean stem heights of 10 year-old miombo trees under plantation conditions in Copperbelt Province, Zambia	137
Figure 5.5	Stem height growth of three miombo trees under plantation conditions in Copperbelt Province, Zambia	137
Figure 5.6	Seedling emergence in *Acacia polycantha*, *Afzelia quanzensis*, *Tamarindus indica*, *Bauhinia petersiana* in untreated soil and earth-kiln burnt	141
Figure 5.7	River basins and forest reserves in Zambia	144
Figure 5.8	Size distribution of forest reserves and national parks in Zambia	146

List of text boxes

Box 4.1	Comparing differences between random and systematic sampling in a miombo woodland inventory	76
Box 4.2	Assessing litter biomass in Mwekera coppice miombo plots using microplots	82
Box 4.3	Developing biomass equations using correlation and regression analysis	89
Box 4.4	Results of biomass equations using correlation and regression analysis (continued from Box 4.3)	89
Box 4.5	Calculation of indices of species diversity	101
Box 4.6	Calculation of similarity coefficients for binary data	103
Box 4.7	Problems associated with the critical population density concept for chitemene shifting cultivation in Zambia	110

Box 4.8	Problems of determining a spatial pattern of deforestation in Zambia	116
Box 4.9	Using change in relative importance to assess response of miombo woodland trees to fire management	117
Box 5.1	Tree conservation in fields and villages in Zambia	135

List of equations

Standard deviation (1)	75
Sampling plot selection (2)	79
Soil bulk density (3)	86
Water saturation point (4)	87
Basal area (5)	88
Solid volume (6)	88
Solid volume regression (7)	91
Biomass equation (based on girth) (8)	91
Biomass equation (based on height and girth) (9)	91
Simpson's dominance index (10)	98
Simpson's diversity index (11)	99
Shannon-Wiener diversity index (12)	99
Index of species evenness (13)	99
Jaccard's coefficient of similarity (14)	100
Sorensen's coefficient of similarity (15)	100
Simple matching coefficcient of similarity (16)	101
Baroni-Urbani and Buser coefficient of similarity (17)	101
Percentage coefficient of similarity (18)	102
Critical population density (19)	108
Fallow ratio (20)	108
Percentage cultivable land (21)	109
Charcoal yield (24)	111
Wood use for charcoal production (25)	114
Statistic for sample proportions (26)	118
Fire efficiency (27)	120
Herbivore pressure (28)	123
Litter mass loss (29)	124

Foreword

THE INTEREST IN indigenous forest management has grown since the Southern African Development Community regional workshop on indigenous forest management held at Victoria Falls, Zimbabwe, in 1989. This was followed by another workshop with the same theme and venue organized by the Zimbabwe Forestry Commission and SAREC in 1992. These workshops recognized the inadequacy of technical capacity to manage indigenous forests in southern Africa. To a large extent, this inadequacy is preventing tropical dry forests in the region from being managed sustainably. Miombo is the most extensive of these regional forests.

The idea to produce a Handbook of Miombo Ecology and Management emerged during the implementation of a research study to assess responses of miombo woodland to harvesting and management, one of the 11 projects under the Charcoal Utilization Programme funded during 1989–93 by the Stockholm Environment Institute (SEI) in co-operation with the Department of Energy in the Ministry of Energy and Water Development. Both the scarcity of appropriate training materials, especially relevant literature, and inadequate knowledge about the ecology and silviculture of indigenous forests have contributed to a lack of technical and professional skills in forest management. The Miombo handbook proposal was endorsed at a national workshop held at Siavonga, Zambia, in May 1993.

The Miombo handbook idea was further developed during discussions with Professor Hans Egneus and Anders Ellegård of BioQuest, Sweden, who were instrumental in securing funding for the project from SEI. Given the scarcity of literature on miombo ecology, it was decided that the book should be based on research results from the study on responses of miombo to harvesting and management and literature review. A preliminary version of this book was printed and used as education material at a training course for Forest Department officers at the Mwekera Forestry College in April, 1995.

The handbook has five chapters. The first part of chapter 1 is an introduction to the indigenous vegetation types in the miombo region, commonly referred to as the Zambezian phytoregion. The second part of the chapter is a description of miombo ecology.

Chapter 2 is a description of the various major uses of the miombo ecosystem. Chapter 3 is a biogeography of fire, one of the principal ecological factors and a significant management problem, in the miombo ecosystem and a description of adaptations in different plant forms. Methods of carrying out management inventories and assessments of various uses of miombo are presented in chapter 4, and management guidelines are presented in chapter 5.

The little that is known about the ecology of miombo is based on a few site-specific studies, often conducted without replication. Validation of such results is therefore difficult and generalizations from such studies can be misleading. Nevertheless, generalizations are useful in presenting a thesis of the current ecological knowledge about miombo, especially for the non-professional. The handbook is based on this approach, and the reader should bear this weakness in mind. For this reason methods and guidelines presented in chapters 4 and 5 should be treated as generic, and their applicability may require modification to suit local conditions. The intentions of this book are to broaden the user's knowledge of the ecology, value and management of the miombo ecosystem.

EMMANUEL N. CHIDUMAYO
Biology Department
University of Zambia
P.O. Box 32379
Lusaka
Zambia

Acknowledgements

The preparation, production and testing of this handbook at a training course held in April 1995, and its subsequent revision, were funded by the Stockholm Environment Institute. I wish to thank Dr Anders Ellegård of BioQuest for his support throughout the handbook development process and Professor Hans Egnéus for his critical comments on the first draft which shaped the form of the final handbook. I am also grateful to the Department of Energy in the Ministry of Energy and Water Development in the Government of Zambia for the logistical support, and Zambia Forestry College in Kitwe, Zambia, for organizing the training course that tested the suitability of the draft handbook.

Abbreviations

AG	above ground
BA	basal area
BABH	basal area at breast height (1.3 m AG)
BD	basal diameter
BG	below ground
bh	breast height (1.3 m AG)
CPD	critical species diversity
dbh	diameter at bh
cm	centimetre
ETP	evapotranspiration
g	gram
gbh	girth at bh
GMA	game management area
ha	hectare
h	hour
kg	kilogram
km	kilometre
LAI	leaf area index
m	metre
M	million
MC	moisture content
NDVI	normalized difference vegetation index
OD	oven-dry
R	rainfall
r	correlation coefficient
SD	standard deviation
SE	standard error = $SD/\sqrt{n}$
sp	species (singular)
spp	species (plural)
sq.	square
SV	solid volume of wood
STV	stacked volume of wood
t	tonne (1000 kg)
yr	year
°	degrees
°C	degrees Celsius

CHAPTER 1
Introduction

The Zambezian phytoregion

THE ZAMBEZIAN PHYTOREGION extends over 10 countries in central and southern Africa lying between latitudes 3° and 26° south with a total area of 377 million ha (White 1983: Figure 1.1). The regional climate is continental with summer rainfall from November to April and a seasonal variation in mean temperature of 18–24 °C. Annual rainfall ranges from 500–1500 mm which decreases from north to south. The dominant soils in the Zambezian phytoregion are Rhodic and Haplic Nitosols and Chromic Xerosols with Calci-Chromic Cambisols and Pellic Vertisols in some places.

The Zambezian phytoregion has about 8500 species of which about 4600 are endemic (White 1983). Among the endemic taxa are *Bolusanthus*, *Cleistochlamys*, *Colophospermum*, *Diplorhynchus*, *Pseudolachnostylis* and *Viridivia*. The phytoregion is also the centre of diversity of the genera *Brachystegia* and *Monotes* which, together with the endemic *Diplorhynchus* and *Pseudolachnostylis*, are also found in miombo woodland. Although miombo is a very extensive vegetation, there are other important vegetation types in the Zambezian phytoregion. These are briefly described below.

Dry forest
Dry forest occurs on deep, freely drained soils with enough supply of moisture in lower horizons during the dry season. There are two types: evergreen and deciduous. Dry evergreen forest is confined to the wetter northern parts of the phytoregion with a mean annual rainfall of more than 1200 mm, except on kalahari sand where it occurs in areas with 900–1200 mm. Although variable in floristic composition, the dominant trees include *Berlinia giorgii*, *Cryptosepalum pseudotaxus*, *Daniellia alsteeniana*, *Entandrophragma delevoyi*, *Marquesia acuminata*, *M. macroura*, *Parinari excelsa* and *Syzygium guineense afromontanum*. Apparently dry evergreen forest has been destroyed by cultivation and fire and only tiny and disturbed fragments remain, usually amidst secondary grassland and wooded grassland (Fanshawe 1971; White 1983).

In areas with 600–900 mm annual rainfall, but with deep sandy soils or lateral seepage water, dry evergreen forest is replaced by deciduous forest and scrub. The most extensive deciduous forest is the *Baikiaea* forest on

Figure 1.1 *The Zambezian phytoregion (shaded area) of central and southern Africa*

Figure 1.2 *Central and southern African countries with miombo vegetation. 1 = Lubumbashi, 2 = Lusaka, 3 = Harare*

kalahari sand in the southern part of the upper Zambezi basin in northeastern Botswana, south-western Zambia and north-western Zimbabwe. *Baikiaea plurijuga* is the dominant tree, with *Pterocarpus antunesii* as a subdominant and *Ricinodendron rautanenii* locally dominant. *P. antunesii* is common in other types of dry deciduous forest in the valleys of middle and lower Zambezi and its tributaries in which *B. plurijuga* is absent. In the dry scrub forest of western Angola, *Adansonia digitata, Sterculia setigera* and *Euphorbia conspicua* occur as emergents.

Swamp and riparian forest
In the wetter parts of the Zambezian phytoregion, swamp and riparian forest occur around and along permanent watercourses. The floristic composition varies with substrate, climate and the extent of flooding. However, the dominant trees include *Mitragyna stipulosa, Syzygium owariense, Xylopia aethiopica, X. rubescens* and *Uapaca guineensis. Ficus congensis, Gardenia imperialis, Ilex mitis, Syzygium cordatum* and *Raphia farinifera* also occur, but their presence in the absence of the other emergents is said to signify regression from swamp forest (Fanshawe 1971). Understorey species include *Aporrhiza nitida, Garcinia smeathmannii* and *G. imperialis*.

Woodland
Woodland is the most widespread and characteristic vegetation of the Zambezian phytoregion. In many places it is the climax vegetation, but in others it is secondary or greatly modified by cultivation and fire. The four distinct woodland types are miombo, mopane, munga and chipya.

Mopane woodland and scrub. Mopane is a valley vegetation type and is most extensive in parts of the Zambezi, Luangwa, Limpopo, Shashi and Sabi valleys, as well as in the Makarikari and Okavango basins in Botswana. It also covers extensive areas in Namibia and south-west Angola. The mopane habitat is characterized by low rainfall and high temperatures with a variety of soils, including sodic soils. *Colophospermum mopane* is the dominant species which often forms pure stands. In the Zambezi and Luangwa valleys the associates of *C. mopane* are *Acacia nigrescens, Adansonia digitata, Combretum imberbe, Sclerocarya caffra* and *Kirkia acuminata*. In the western part of the mopane range in Botswana, Angola and Namibia, the main associates of *C. mopane* are *Acacia erubescens, A. kirkii, Balanites angolensis,* and species of *Boscia, Combretum, Terminalia* and *Commiphora. C. mopane* is resinous and therefore easily burns, especially when the bark has been destroyed by fire.

Munga woodland and scrub. Munga or undifferentiated woodland (Fanshawe 1971, White 1983) is an open deciduous vegetation with scattered or grouped emergents dominated by species of *Acacia, Combretum and Terminalia*. Other characteristic genera are *Adansonia, Afzelia, Albizia, Burkea* and *Borassus* and *Hyphaene* palms. Munga woodland occurs on flat topography in drier parts of the Zambezian phytoregion and on lacustrine or riverine alluvial soils with a high base exchange capacity. White (1983) has suggested that in riparian habitats in the drier parts of the phytoregion, munga woodland represents a degraded form or replacement of riparian forest.

Chipya woodland. The term chipya is applied to wooded grassland in areas with more than 1000 mm annual rainfall on the central African plateau. Chipya woodland is most extensive on alluvial soils of lake basins, especially the Lake Bangweulu system with its associated rivers. Chipya vegetation consists of a complex mosaic representing different stages of degradation and re-establishment of dry evergreen forest represented by small patches of evergreen thickets (mateshi) and scattered *Entandrophragma delevoyi* emergents. The herb layer is dense and tall (2–3 m) with *Hyparrhenia* and *Andropogon* as characteristic grasses and *Afromomum biauriculatum*, *Pteridium aquilinum* and *Smilax kraussiana* ferns. Chipya trees are said to be fire resistant (Lawton 1978), and include *Afzelia quanzensis, Albizia antunesiana, Burkea africana, Erythrophleum africanum, Parinari curatellifolia, Pericopsis angolensis* and *Pterocarpus angolensis.*

Thicket
Various types of thickets occur scattered throughout the Zambezian phytoregion. These occupy a variety of habitats. The itigi thicket of central Tanzania and north-eastern Zambia is a dense deciduous vegetation on specialized soils that are well aerated and watered during the rainy season. The shrubs are interlaced overhead to form a thick continuous canopy at 3–5 m height dominated by *Baphia, Burttia, Combretum* and *Grewia* species with scattered emergents, such as *Albizia petersiana* or *Craibia brevicaudata.*

The Pemba thicket occurs on shallow transitional soils (Inceptisols) on the edge of the miombo-dominated plateau in southern central Zambia. This is 6–7 m tall impenetrable deciduous thicket in which *Acalypha chirindica, Aeschynomene trigonocarpa, Byrsocarpus orientalis, Canthium burtii, Cassipourea gossweileri* and *Combretum celastroides* are abundant. Emergent trees include *Combretum collinum, Lannea discolor, Parinari curatellifolia, Pericopsis angolensis, Peltophorum africanum, Pterocarpus angolensis* and *P. rotundifolius.*

Termitaria thicket occurs on termite mounds in different vegetation types. The characteristic genera are *Acacia, Albizia, Asparagus, Canthium, Combretum, Commiphora, Cassia, Ficus, Grewia, Popowia, Sanseviera, Ximenia* and *Ziziphus.*

Rupicolous thickets occur on rocky outcrops, especially in the drier parts of the Zambezian phytoregion. The flora is dominated by *Bauhinia petersiana, Canthium* spp., *Cassia abbreviata, Commiphora mossambicensis, Euclea natalensis, Euphorbia candelabrum, Ficus ingens, F. sonderi, Hippocratea indica, Lannea discolor, Landophia parvifolia* and *Strychnos potatorum.*

Grassland
Grasslands in the Zambezian phytoregion are either associated with seasonally waterlogged soil or toxic soils. Seasonally waterlogged habitats

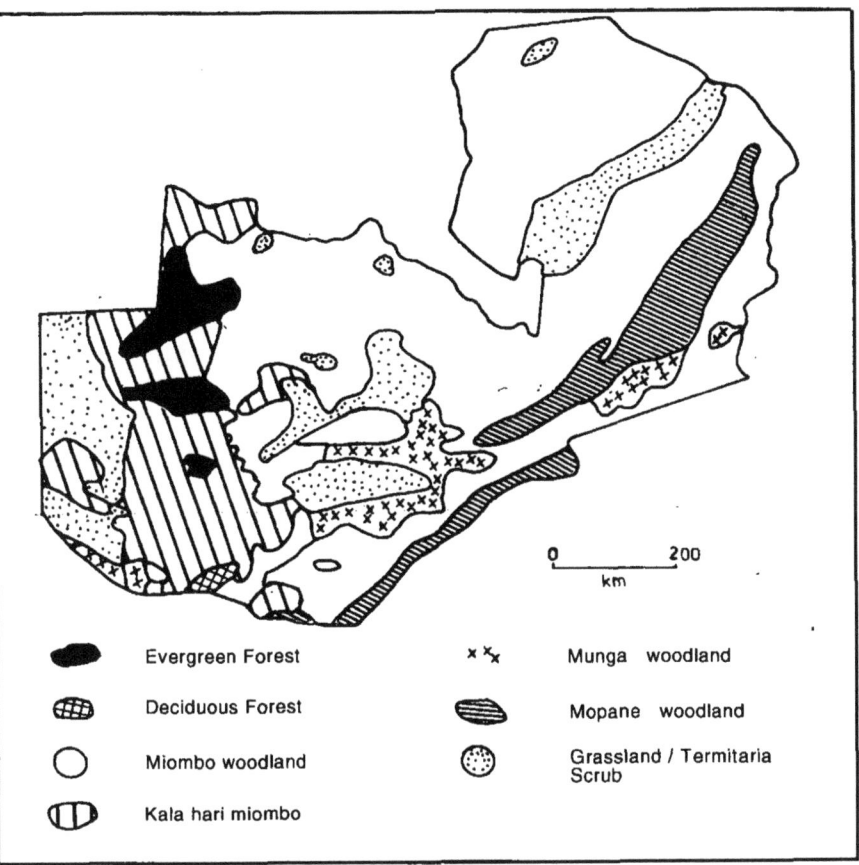

Figure 1.3 *Generalized vegetation map of Zambia*

occur in shallow depressions (dambos) which form the headwater reaches of watercourses and on flood plains in river valleys and basins.

Dambo grassland occupies up to 20 per cent of the central African plateau above 1200 m altitude (White 1983). The dambo may be perennially or seasonally flooded and the soil tends to be acid. Dambo grasslands are frequently burnt and *Loudetia simplex* is the most characteristic grass. Other common grasses are *Andropogon schirensis, Hyparrhenia* spp., *Miscanthus teretifolius, Monocymbium ceresiiforme, Themeda triandra* and *Trachypogon spicatus*. Sedges may be common and even dominant in some dambos. Among the common dambo sedges are *Ascolepsis* spp., *Bulbostylis cinnamomea, Cyperus* spp., *Kyllinga erecta, Mariscus deciduus, Scirpus microcephalus* and *Scleria bulbifera*. A variety of forms may also be conspicuous in dambo grasslands.

Flood plain (wetland) grassland occur on valleys of the larger rivers on alluvium. The principal grasses are *Acroceras macrum, Echinochloa pyramidalis, E. scabra, Leersia hexandra, Oryza longistaminata, Panicum repens, Paspalum sacrobiculatum, Sacciolepsis africana* and *Vossia cuspidata*.

The ecology of miombo woodland

Miombo woodland is the most extensive vegetation type in the Zambezian phytoregion. It is found in seven central and southern African countries (Figure 1.2) and covers 270 million ha (Millington *et al.* 1986) which is about 70 per cent of the Zambezian phytoregion. Miombo is characterized by the presence of leguminous trees of the genera *Brachystegia, Isoberlinia* and *Julbernardia*. Miombo is the dominant vegetation in Zambia and covers 53 per cent of the country (Figure 1.3; Table 1.1) with an estimated woody plant flora of 650 species (Fanshawe 1971).

Table 1.1 *Distribution of land cover types in Zambia by province (km^2)*

				PROVINCE						ZAMBIA
		Central	Copper-belt	Luapula	Lusaka	Eastern	North-ern	North-western	South-ern	West-ern	
FOREST											
	Dry evergreen	0	0	0	0	0	380	8 580	0	1 100	16 060
	Deciduous	620	0	0	0	0	0	440	1 940	3 830	6 830
	Montane	0	0	0	0	0	0	40	0	0	40
	Swamp	0	20	0	0	0	20	1 490	0	0	1 530
	Riparian	0	0	10	30	640	10	30	10	80	810
	Plantation	0	500	0	0	0	0	0	0	0	500
	Chipya	1 500	1 230	4 580	0	30	6 470	510	20	1 220	15 560
	Thicket	0	0	150	420	80	1 250	0	0	0	1900
MIOMBO WOODLAND											
	Wet	9 410	20 290	25 730	0	0	81 010	51 560	0	3 850	191 840
	Dry	40 750	0	0	13 470	34 210	2 360	0	28 830	0	119 620
	Kalahari	310	170	0	0	0	0	22 000	7 240	55 740	85 460
SAVANNA WOODLAND											
	Mopane	5 130	0	0	0	14 330	8 820	150	8 880	1 390	38 700
	Munga	7 100	10	10	4 090	8 050	710	240	9 130	3 260	32 600
	Termitaria	3 600	560	340	60	640	5 770	4 590	1 720	6 980	24 260
GRASSLAND											
	Wetland	15 860	5 280	10 940	1 220	2 920	22 710	23 620	16 770	30 710	130 030
	Dambo	9 870	3 250	4410	2 580	8 200	15 250	12 570	8 260	11 930	76 320
OPEN WATER		240	20	4370	20	0	3 070	0	2 470	310	10 500
TOTAL		94 390	31 330	50 540	21 890	69 100	147 830	125 810	85 270	126 400	752 560

Based on Forest Department/FAO (1986) and Chidumayo (1994a).

Climate

The climate in the miombo region is characterized by an alternation of dry and wet seasons (Figure 1.4). The average annual rainfall of 600–1500 mm is distributed from November to March or April. Three seasons based on temperature and rainfall are recognized in Zambia (Table 1.2): hot, dry (September–November), hot, wet (December–March) and cool, dry (April–August).

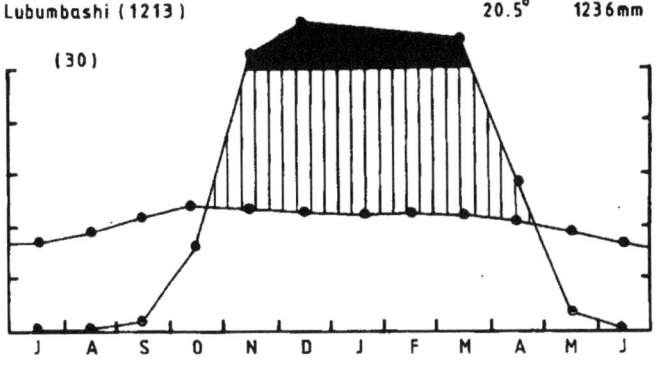
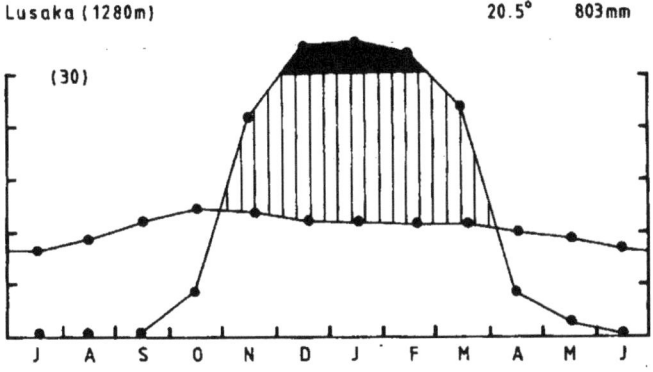
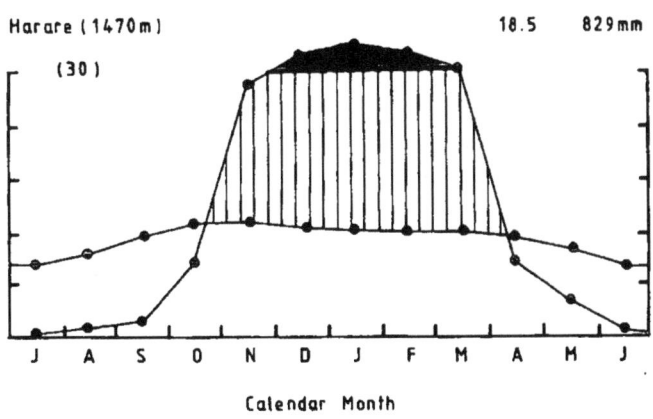

Figure 1.4 *Climate diagrams for three stations in miombo region*

Miombo trees are frost sensitive and do not tolerate absolute minimum temperatures of less than –4 °C and this may be the significant factor that determines the southern limit of miombo distribution in the Zambezian phytoregion (Werger and Coetzee 1978). However, within the miombo range, frost may occur in low-lying areas and in dambos during the cool dry season. The high temperatures in the hot dry season (Table 1.2) are responsible for the high rates of potential evapotranspiration (ETP) in relation to rainfall (R). The aridity ratio (R/ETP) in Zambian miombo ranges from 0.8–1.7 (Chidumayo 1987a).

Table 1.2 Climatic seasons at Lusaka, central Zambia

Season	Duration	Mean rainfall (mm)	Mean temperature range (°C)	
			Minimum	Maximum
Hot dry	Sep–Nov	95	14–18	29–32
Hot wet	Dec–Mar	670	16–18	25–27
Cool dry	Apr–Aug	24	9–14	23–26

White (1983) divided miombo into dry (< 1000 mm mean annual rainfall) and wet (> 1000 mm) types. Chidumayo (1987a) further subdivided each of the two miombo types in Zambia into subtypes.

Wet miombo has two subtypes. The northern wet miombo consists of *Brachystegia (B. spiciformis–B. utilis)* woodlands with *Julbernardia paniculata* and *Parinari curatellifolia* as common canopy co-dominants, and *Monotes africanus*, *Syzygium guineense macrocarpum* and *Uapaca* spp. as common understorey taxa. This type is found in Northern and Luapula Provinces and northern parts of Central Province (for province location, see Appendix 1). The north-western wet miombo consists of *Brachystegia (B. spiciformis, B. longifolia)* woodlands with *Isoberlinia angolensis* and *J. paniculata* as common canopy co-dominants, and *Anisophyllea boehmii*, *Diplorhynchus condylocarpon*, *Syzygium guineense macrocarpum* and *Uapaca* spp. as common understorey trees, and is found in Copperbelt and North-western Provinces and the northern parts of Western Province.

Dry miombo has three subtypes. The central dry miombo consists of *Brachystegia (B. boehmii–B. spiciformis–B. utilis)* woodlands with *Julbernardia globiflora* as a common canopy co-dominant and *D. condylocarpon*, *Lannea* spp., *Ochna* spp. and *Pseudolachnostylis maprounefolia* as common understorey trees and is found in Central Province and parts of Southern Province. The eastern dry miombo consists of *Brachystegia manga–Julbernardia* woodlands with *Diospyros* spp., *D. condylocarpon*, *Ochna* spp. and *P. maprounefolia* as common understorey trees, and is found in Eastern Province and the adjoining parts of Central and Northern Provinces. The western dry miombo consists of *Brachystegia spiciformis–J. paniculata* woodlands with *Burkea africana* as a common canopy co-dominant and *D. condylocarpon* as a common understorey species, and is found in

Table 1.3 *Mean species diversity in miombo subtypes in Zambia*

	Wet miombo		Dry miombo		
	Northern	North-western	Central	Eastern	Western
Species 0.1 ha^{-1}	18.30	18.78	14.79	16.65	12.29
Species 0.4 ha^{-1}	29.68	30.33	21.90	24.50	19.14
Simpson's Index of dominance	0.106	0.110	0.135	0.128	0.135

Based on Chidumayo (1987a)

Western Province and parts of Southern Province. This dry miombo type has also been referred to as kalahari miombo (White 1983) or kalahari sand woodland (Fanshawe 1971). Species diversity in the five miombo subtypes is slightly different, but the difference is largest between dry and wet miombo (Table 1.3).

Soils

Miombo soils are of eluvial origin on basement quartzites, schists and granitic rocks. These soils cover most of the central African plateau, which is an end-tertiary pediplain (Cole 1963; Fanshawe 1971). The soil texture is sandy loam, sand clay loam and sand clay. As a result of the eluviation process, the clay content increases with depth (Table 1.4). Soil colour varies from shades of brown in the top soil (0–30 cm), to reddish and orange in the bottom soil on well and poorly drained sites, respectively (Chidumayo 1993a).

Table 1.4 *Average textural composition of dry miombo soils in central Zambia*

Soil depth (cm)	Clay %	Silt %	Sand %
0–10	11	12	77
11–30	11	17	72
31–60	22	10	69
61–100	31	9	60

Based on Chidumayo (1993a)

In Zambia miombo soils are classified as Oxisols, Ultisols and Alfisols. Oxisols and Ultisols are more acidic (pH 4–5) and are covered by wet miombo, whereas Alfisols (pH 5–6) are covered by dry miombo. Soil acidity also increases with depth. For example, at four central dry miombo sites in Zambia average pH decreased from 5.6 at 0–10 cm soil depth to 5.2 at 11–60 cm and 4.9 at 11–150 cm depth (Chidumayo 1993a). Miombo vegetation is absent on soils that are alkaline (pH > 7) and those that are waterlogged. Under these edaphic conditions, miombo gives way to mopane or munga woodland and swamp or riparian forest.

There is considerable variation in soil depth. Plateau soils may be very deep (Celander 1983), while on hills and escarpments soils are extremely

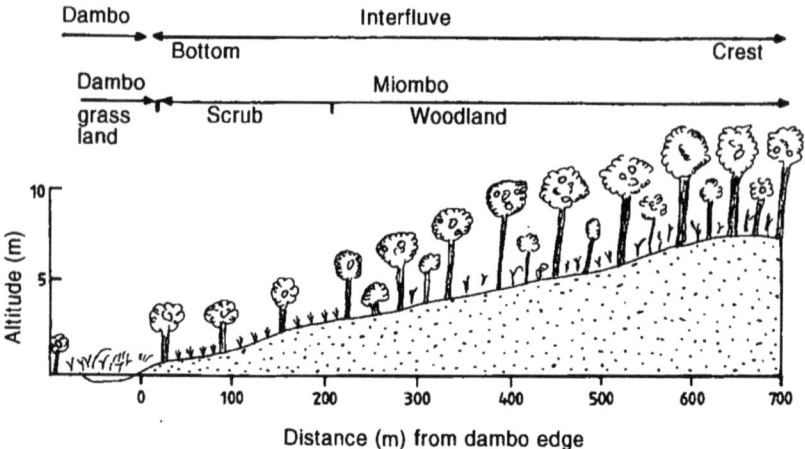

Figure 1.5 *Vegetation catena in miombo from dambo to interfluve crest*

shallow with surface rock boulders. Miombo shows a catenary sequence in relation to topographic gradient, from well-developed plateau woodland to poorly developed hill and escarpment woodland and scrub on inselbergs. Scrub miombo is also associated with impeded drainage on dambo edges (Figure 1.5) and where a lateritic pan is close to the surface. The sharp miombo boundary at the edge of dambos is also reinforced by the frequent occurrence of frost to which miombo species are intolerant. Some miombo trees are indicator species (Storrs 1982). *B. boehmii* and *Parinari curatellifolia* are indicators of shallow soils with partial waterlogging or high water table. *D. condylocarpon* and *P. maprounefolia* often indicate the presence of copper or nickel. Other species, such as *Brachystegia microphylla* in wet miombo and *B. glaucescens* in dry miombo prefer rocky habitats (Fanshawe 1971). Edaphic gradients determine the distribution of miombo species and specific preferences were demonstrated by Savory (1962) (Table 1.5). Edaphic changes can occur over short distances and result in complex patterns of species dominance. Hills and escarpments have greater small-scale substrate heterogeneity and more complex sequences of species dominance than plateau landscapes (Figures 1.6a and b). The bulk densities of miombo soils range from 1.2–1.5 t m^{-3} with lower values (1.2–1.3) for top soil (Chidumayo 1993a).

Top-soil moisture content (MC) is very variable because of seasonality in rainfall (Figure 1.7). Average soil MC at four dry miombo sites in central Zambia ranged from 1.5–5.0 per cent in the dry season to 9.5–15.5 per cent in the rainy season (Chidumayo 1993a).

Bottom soil MC at 50–70 cm depth ranged from 10 per cent during July–October to 30 per cent in the middle of the rainy season at a Zimbabwe dry miombo site (Ernst and Walker 1973).

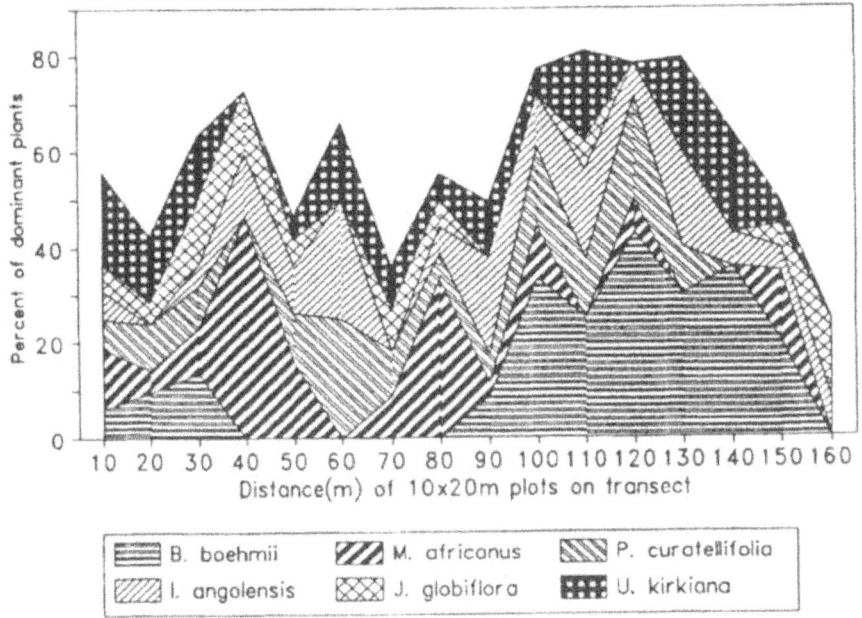

Figure 1.6a *Spatial successions of dominant species along a 160 m hillside transect in central dry miombo*

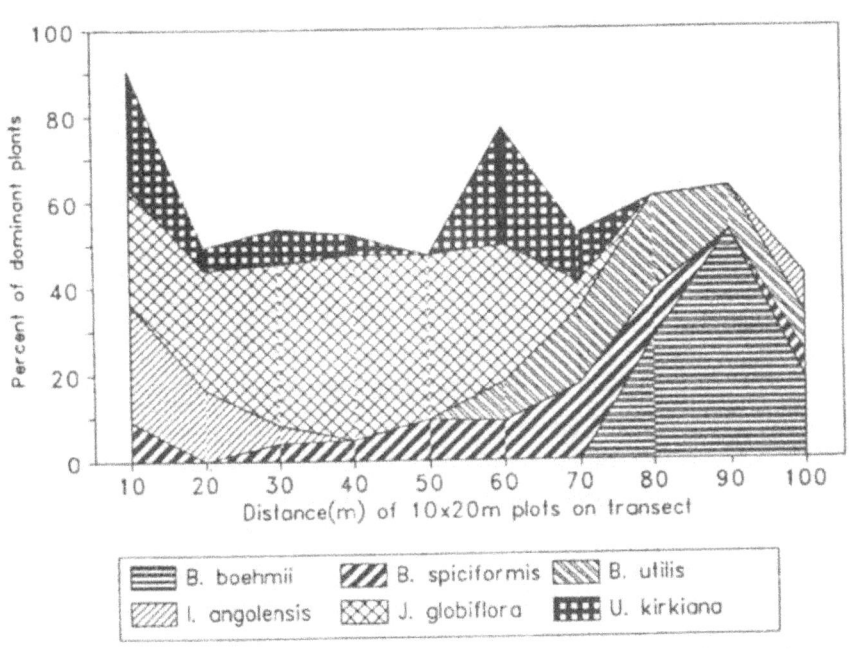

Figure 1.6b *Spatial successions of dominant species along a 100 m interfluve crest transect in central dry miombo*

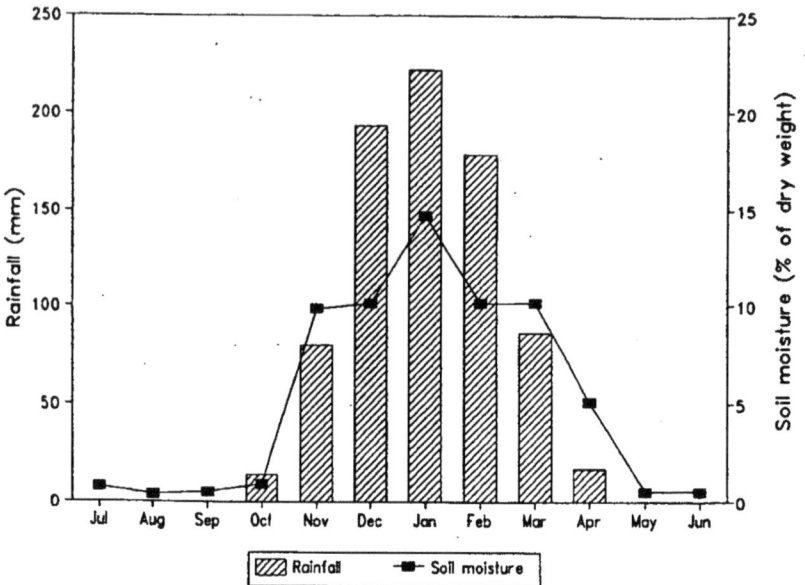

Figure 1.7 Seasonality in rainfall and top (0–30 cm) soil moisture content in dry miombo

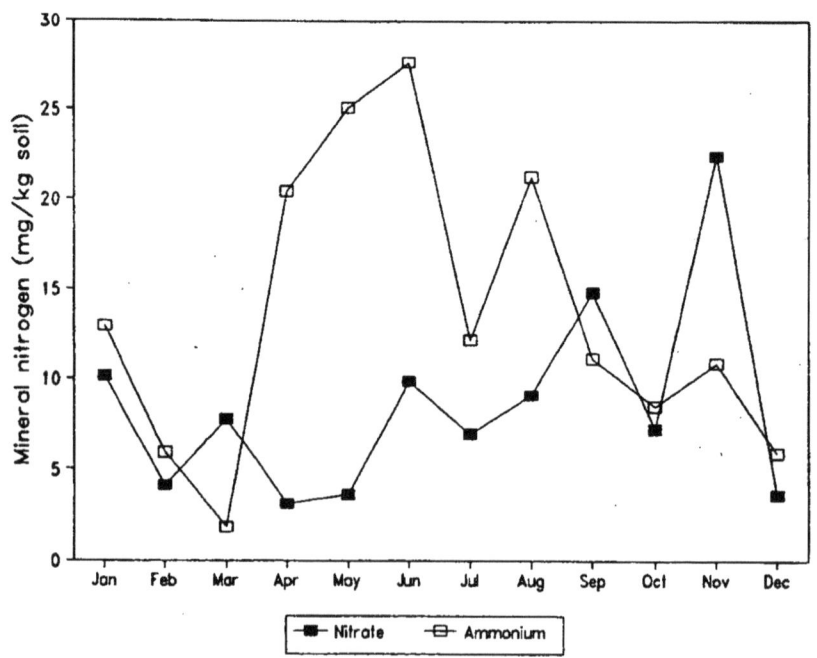

Figure 1.8 Monthly variation in ammonium and nitrate concentration in dry miombo top soil

Table 1.5 *Edaphic preferences by* Brachystegia *species in the Zambian Copperbelt*

Species	Soil preference
Brachystegia boehmii	Clay loam
Brachystegia floribunda	Heavy-textured soil
Brachystegia longifolia	Deep sandy soil
Brachystegia spiciformis	Deep well drained soil
Brachystegia utilis	Deep loams

Based on Savory (1962)

Miombo soils have low concentrations of organic matter, macronutrients and exchangeable bases. Top-soil organic matter content is about 1.0 per cent and 2.0 per cent in wet and dry miombo, respectively (Chidumayo 1993a; Stromgaard 1984) and this decreases with depth (Table 1.6).

Table 1.6 *Soil organic matter content (% of dry weight) in dry and wet miombo*

Soil depth	Dry miombo	Wet miombo
0–10	1.6	1.1
11–30	0.7	0.5
31–60	0.8	–
61–150	0.6	0.2

Based on Chidumayo (1993a) for dry miombo and Stromgaard (1984) for wet miombo

Total soil nitrogen (N) content is less than 0.1 per cent (Chidumayo 1993a; Stromgaard 1989), while mineral N varies throughout the year and is dominated by ammonium N. In central dry miombo, the ammonium N ranges from 2–30 mg kg^{-1} soil while that of nitrate N ranges from 2–20 mg kg^{-1} (Figure 8), with no variation with depth (Chidumayo 1993a).

The concentration of extractable phosphorus (P) in miombo soils varies seasonally and with soil depth (Chidumayo 1993a). The concentration is 5–15 mg kg^{-1} soil in the cool dry season (April–August) and is higher (15–45 mg kg^{-1}) during the other seasons. At four central dry miombo sites in Zambia, the average concentration of extractable P decreased from 24 mg kg^{-1} at 0–10 cm depth to 19 at 11–30 cm, 11 at 31–60 cm and 3 mg kg^{-1} at 61–150 cm (Chidumayo 1993a).

Table 1.7 *Average concentration of exchangeable bases (meq 100g^{-1}) in dry and wet miombo top soil*

Exchangeable base	Dry miombo	Wet miombo
Calcium	2.76	0.75
Magnesium	1.82	0.42
Potassium	0.84	0.20
Sodium	0.08	0.06

Based on Chidumayo (1993a) for dry miombo and Stromgaard (1989) for wet miombo

The average concentration of exchangeable bases is higher in dry than wet miombo soils (Table 1.7). There are considerable temporal variations in the concentration of exchangeable potassium (K) and sodium (Na) (Figure 1.9). The concentration of exchangeable K is lowest (< 0.5 meq 100 g^{-1} soil) at the end of the dry season. A similar trend occurs in the concentration of exchangeable Na. Cation exchange capacity in miombo soils is 3.8 meq 100 g^{-1} and 5.4 meq 100 g^{-1} in top soil in dry and wet miombo, respectively (Chidumayo 1993a; Stromgaard 1989).

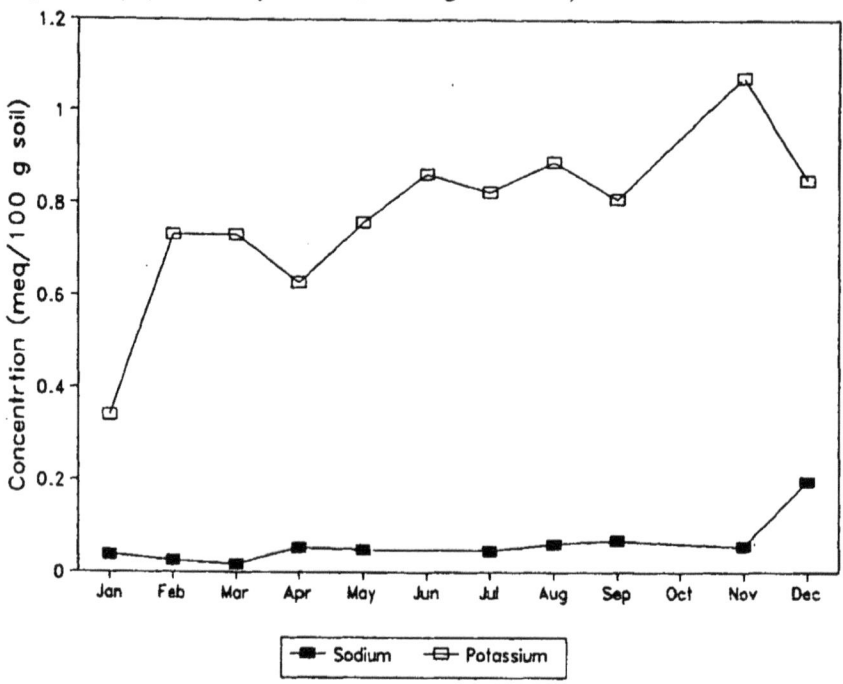

Figure 1.9 *Monthly variation in exchangeable sodium and potassium in dry miombo top soil*

Table 1.8 *Woody plant species diversity in 0.10 ha sample plots in some Zambian forests*

Forest type		Mean species per plot	Species dominance (D)	Species diversity (Dv)
Swamp forest		7.50	0.63	0.37
Acacia woodland		14.00	0.23	0.77
Miombo woodland:				
	kalahari	12.29	0.13	0.87
	dry	15.72	0.13	0.87
	wet	18.54	0.11	0.87
Dry evergreen forest		23.44	0.13	0.87

Based on Chidumayo (unpublished)

Nutrient cycling
Miombo woodland has a higher species diversity than many other vegetation types in the Zambezian phytoregion (Table 1.8). Since miombo woodland grows on poor soils, high species diversity in miombo probably increases the probability for the coexistence of different strategies for nutrient retention. This hypothesis is supported by the differences in nutrient concentrations in plant tissues among miombo woodland species (Table 1.9). The N content of N-fixing species in miombo is 2.5 times that of non-N-fixing species, while P content in *Brachystegia, Julbernardia* and *Isoberlinia* species is 3.3 times that of N-fixing species (Chidumayo 1992a). These differential nutrient concentrations in plant tissues ensure the storage of adequate stocks of a variety of nutrients in plant biomass at community level. High species diversity in miombo may therefore be important to the maintenance of nutrient cycling and ecological stability.

Table 1.9 *Differential nutrient concentration (% of dry weight) in plant tissues of different plants in miombo woodland*

Plant class	Plant tissue	Nitrogen	Phosphorus	Potassium
Grasses	Stem + leaves	0.60	0.8	2.51
	Roots	0.44	0.10	1.21
Other herbs	Stem + leaves	1.35	0.44	4.01
	Roots	0.60	0.19	1.97
Woody plants	Leaves	1.40	0.25	1.63
	Stems	0.62	0.25	1.55
	Roots	0.54	0.11	1.29

Based on Chidumayo (1993a)

Nutrient leaching in wet miombo is probably more significant than in dry miombo. It is therefore probable also that the higher species richness in wet miombo (Table 1.8) may represent more diverse mechanisms for conserving and recycling nutrients in order to counteract the high potential for nutrient loss that is associated with high rainfall. There is a strong internal recycling of limiting nutrients within miombo trees. About 30 per cent of foliar N and 80 per cent of foliar P are withdrawn at the end of the growing period during February to April, and a further 30 per cent of N is withdrawn just before

Table 1.10 *Nitrogen and phosphorus concentration (% dry of weight) in foliage and leaf litter of miombo trees*

Biomass type	Period	Nutrient concentration (% of dry weight)	
		Nitrogen	Phosphorus
Foliage	Feb–Apr	1.65	0.65
	May–Aug	1.15	0.13
Litter	Aug–Nov	0.66	0.14

Based on Chidumayo (1993a)

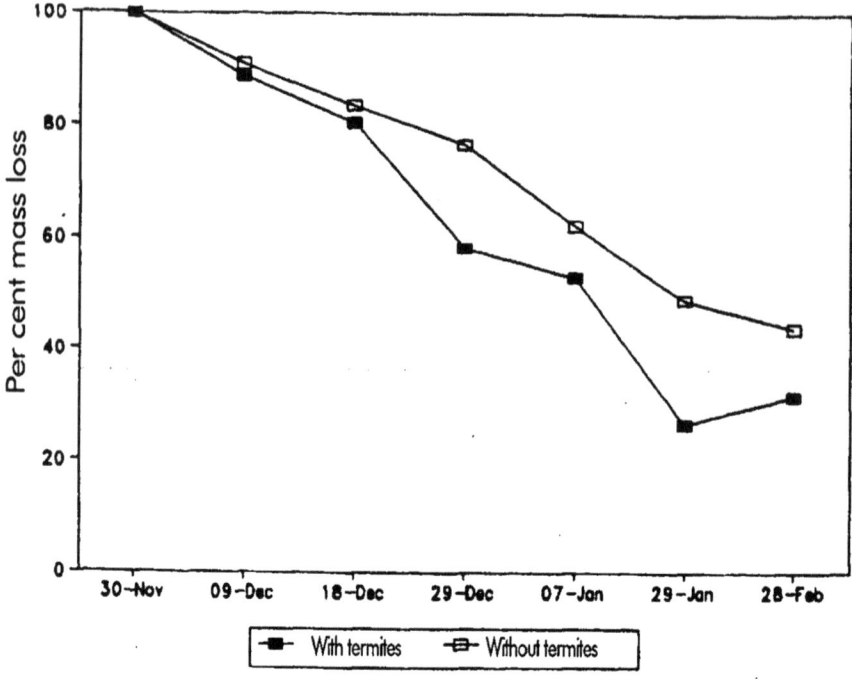

Figure 1.10 *Decomposition of wet miombo tree leaf litter*

leaf fall (Chidumayo 1994b; Table 1.10). This nutrient withdrawal from leaves reduces the loss of valuable nutrients from miombo trees.

Litter input from tree leaf fall and dead AG (above ground) herbaceous plants is a major source of soil organic matter and nutrients. Nutrients are released from litter through decomposition. Decomposition in miombo woodland peaks during the rainy season (Malaisse *et al.* 1975; Figure 1.10; Figure 3.4). The release of nutrients from miombo leaf litter during decomposition in a single rainy season ranges from 60–94 per cent (Table 1.10).

Trees in miombo have deep tap roots and extensive lateral roots (see Figure 4.7) which probably capture nutrients lost from the upper soil layers through leaching. Some of these nutrients are returned to the top soil through leaf fall and litter decomposition. These processes which involve nutrient uptake from the soil, litter production and decomposition ensure a stable but tight cycling of essential nutrients in miombo woodland and are probably the basis for sustainable production in miombo ecosystem, in spite of the low stocks of soil nutrients (Tables 1.6 and 1.7).

Soil conservation
Vegetation cover is one of the major factors that determines soil loss through erosion; others are relief, rainfall and soil type. The seasonal rainfall in the miombo ecosystem comes in the form of heavy storms of short duration (Robinson 1978). This makes water the main agent of soil erosion.

Table 1.11 *Changes in nutrient stocks in tree leaf and grass litter in dry miombo during the rainy season*

	Stocks (kg ha^{-1}) per biomass class			
	Tree leaf litter		Grass litter	
Nutrient	End of dry season	End of rainy season	End of dry season	End of rainy season
Nitrogen	17.2	3.9	4.5	1.9
Phosphorus	3.1	0.3	0.6	0.2
Potassium	26.0	1.5	8.5	1.2
Calcium	13.4	3.2	4.8	0.5
Magnesium	8.5	1.3	2.5	0.4
Sodium	1.4	0.1	0.6	0.1

Based on Chidumayo (1993a)

Plant biomass intercepts raindrops and encourages water infiltration into the soil. This soil-protective function of biomass in miombo operates at three levels: tree canopy, herbaceous plant canopy and surface litter. In wet miombo, tree crowns cover 60–90 per cent of the ground (Araki 1992). Between 12–20 per cent of the coverage is made up of 2–3 crown layers of overlapping trees of similar height or of trees at different heights (Figure 1.11). Tree canopy cover consists mainly of leaves which intercept raindrops. Mean specific leaf area in miombo woodland is 100 cm^2 g^{-1} (Chidumayo unpublished). Leaf biomass ranges from 200–300 g m^{-2} which implies a leaf area index (LAI) of 2–3 m^2 m^{-2}. This means that potentially a m^2 of ground in miombo is covered by 2–3 layers of leaves as a protective cover against soil erosion by raindrops.

Miombo woodland is deciduous and leaf fall peaks during August–October while leaf flush occurs during September–November (Table 3.4). At the onset of the rainy season in November, the new leaf flush intercepts raindrops at the canopy level while the recently fallen leaf litter intercepts raindrops at the soil surface. Miombo trees therefore protect the soil against rain erosion at tree canopy and ground level.

The herbaceous layer in miombo is dominated by plants that regenerate their AG parts annually. Grasses and sedges make up 90 per cent of this herbaceous plant biomass (Chidumayo 1993a). During the dry season the dead herbaceous plant biomass is estimated at 70 and 100 g m^{-2} in dry and wet miombo, respectively (Shea *et al.* 1993). This is the amount of soil protection provided by herbaceous plants in miombo at the onset of the rainy season. Above ground grass production occurs during the rainy season, and cumulative green biomass, as well as soil protection, increase throughout the season (see Figure 4.9). This means that the least protection to the soil by herbaceous plants in miombo occurs at the beginning of the rainy season. Maximum erosion can therefore be expected at the onset of the rainy season.

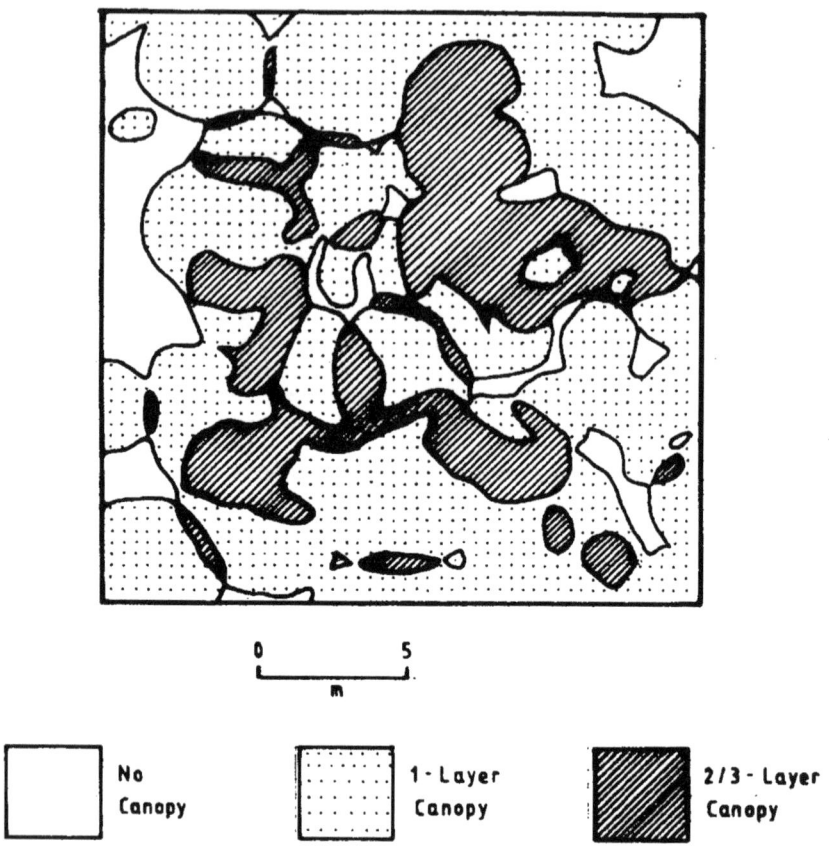

Figure 1.11 *Crown projection in wet miombo (Based on Araki 1992)*

Water use and conservation
The coexistence of perennial woody plants and herbaceous plants in miombo woodland has resulted in a dual water-use pattern.

The productivity of grasses is confined to the rainy season, when top-soil moisture content is > 5 per cent (Figure 1.7). Water use and transpiration loses by herbaceous plants are therefore largely confined to the rainy season, when there is a positive water balance. Consequently, the impact of herbaceous plants on the hydrological cycle in miombo woodland is not as significant as that of woody plants.

Leaf flush marks the onset of the tree growing season in miombo woodland. This occurs 2–3 months before the onset of the rainy season. The greenness of miombo woodland remains high throughout the year, except during August–September (Figure 1.12). This means that evapotranspiration is very high prior to the onset of the rainy season, and this worsens the negative water balance during the hot dry season

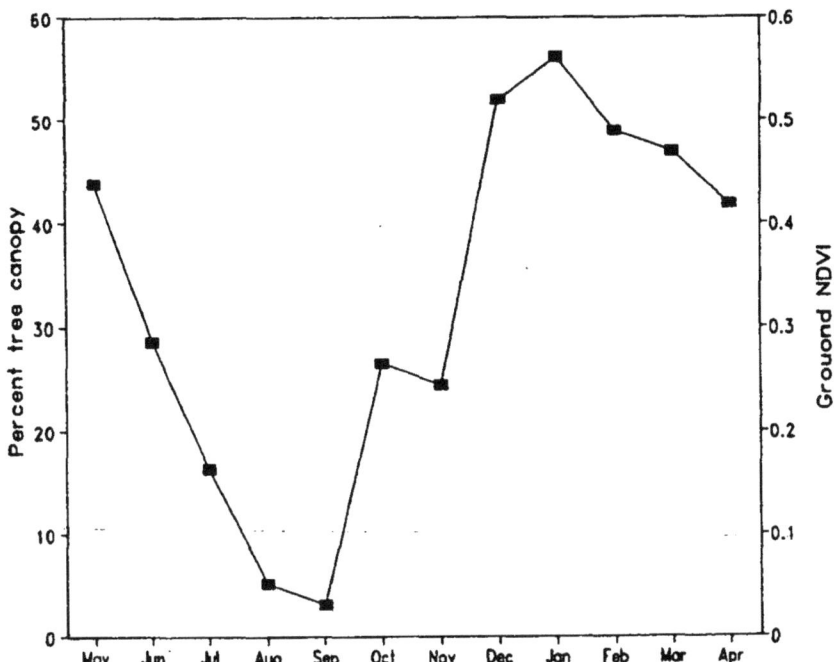

Figure 1.12 *Monthly variation in dry miombo greenness (Based on Fuller, unpublished)*

The mean annual runoff coefficient (as a ratio of mean annual runoff to mean annual rainfall) in Zambia under relatively undisturbed catchment conditions ranges from 0.08–0.24 (Mumeka 1986; Sharma 1984). Stream flow consists of overland runoff and base flow. Overland runoff is associated with the rainy season, while base flow is perennial (Figure 1.13), but each of these contributes equally to river flow in Zambia (Sharma 1984). Monthly river flow hydrographs indicate that ground water is depleted by both base flow and evapotranspiration. Excess evapotranspiration in miombo takes place mainly from August to October as leaf flush and canopy greenness progresses (Figure 1.13). Consequently, the role of miombo woodland in water conservation may be negative (see Chapter 4).

Tree regeneration
Regeneration in miombo can be divided into (i) seed production, dispersal and germination (ii) seedling development and (iii) vegetative regeneration.

Seed production, dispersal and germination Flowering in miombo trees occurs throughout the year with a peak towards the end of the dry season (Figure 1.14) (Chidumayo 1993a; Malaisse 1974), but fruit production varies from year to year, even within the same species (Figure 1.15).

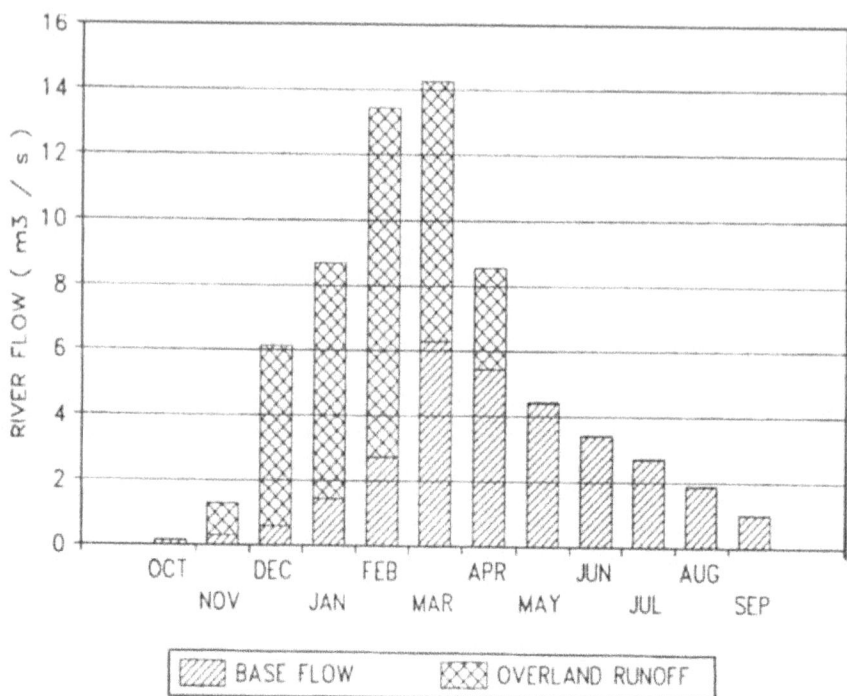

Figure 1.13 *Components of river flow in Zambia (Based on Sharma 1984)*

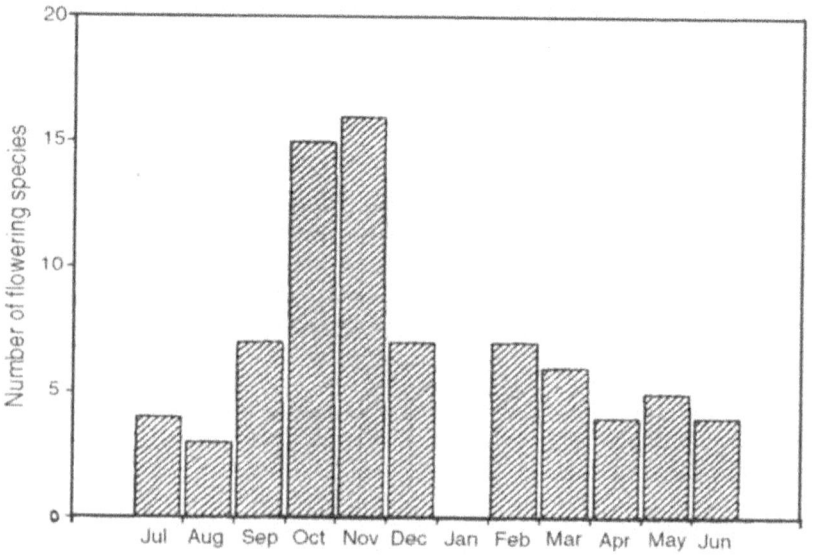

Figure 1.14 *Flowering pattern in dry miombo. None of the species were recorded flowering in January. This may be a sampling artefact.*

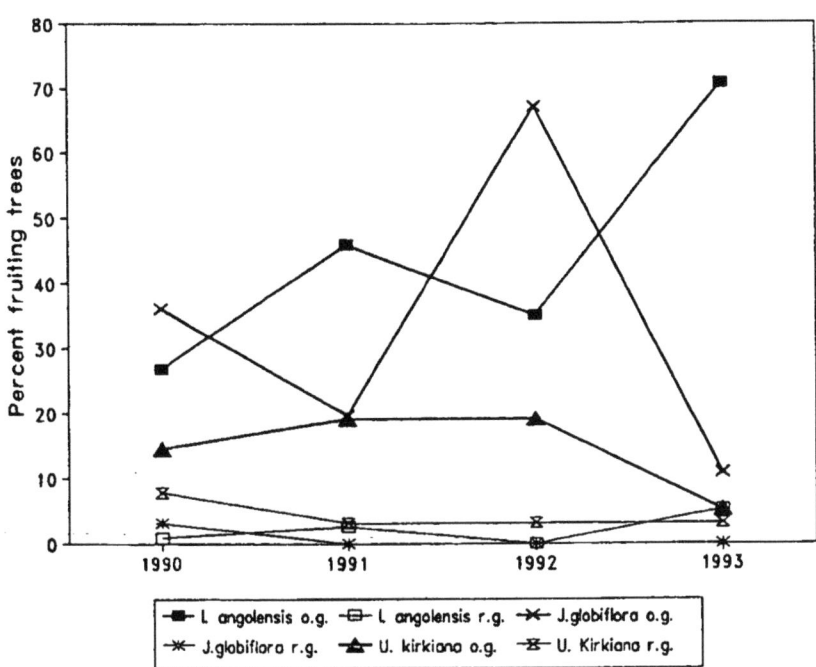

Figure 1.15 *Annual frequency of fruiting in three miombo trees (o.g. = old-growth, r.g. = regrowth)*

Fruiting failure may be caused by a lack of flowering or flower abortion. In *Julbernardia paniculata* and perhaps *Brachystegia* species, flower abortion may be caused by bud infestation by *Curculionid* beetle larvae and imagos (Clauss 1992). Species, such as *Isoberlinia angolensis* and *Julbernardia globiflora*, show great annual fluctuations in the proportion of trees which display masting (fruiting) behaviour.

This behaviour is more pronounced in old-growth than in immature regrowth miombo (Figure 1.15). During mast (fruiting) years, more fruits

Table 1.12 *Diversity of fruit and seed dispersal types among miombo woodland trees and shrubs*

	Fruit type			Seed dispersal agent		
	Fleshy	Dry pod	Other	Wind	Explosive pod	Animal
Canopy	15%	78%	7%	22%	59%	19%
Understorey	48%	24%	28%	42%	4%	54%
Shrub	67%	24%	9%	9%	9%	82%

Based on White (1962)

were produced in old-growth than in non-mast years: *I. angolensis* produced an average of 65 pods per fruiting tree compared to 20 in the non-mast years, while *J. globiflora* produced an average of 846 pods per fruiting tree in a mast year compared to 49 pods in a non-mast year (Chidumayo 1993a). Annual fluctuations in fruit production among miombo trees has also been inferred from annual variations in pod valve litter fall in both dry and wet miombo (Campbell *et al.* 1988; Malaisse *et al.* 1975).

For the majority of miombo trees, the fruit takes at least six months to mature. Exceptions include *J. paniculata* whose fruits mature in less than six months (Figure 1.15). Fruit and seed dispersal is concentrated in the late dry season (August-November), but in a few species, such as *I. angolensis*, seed dispersal continues into the early rainy season (Figure 1.16).

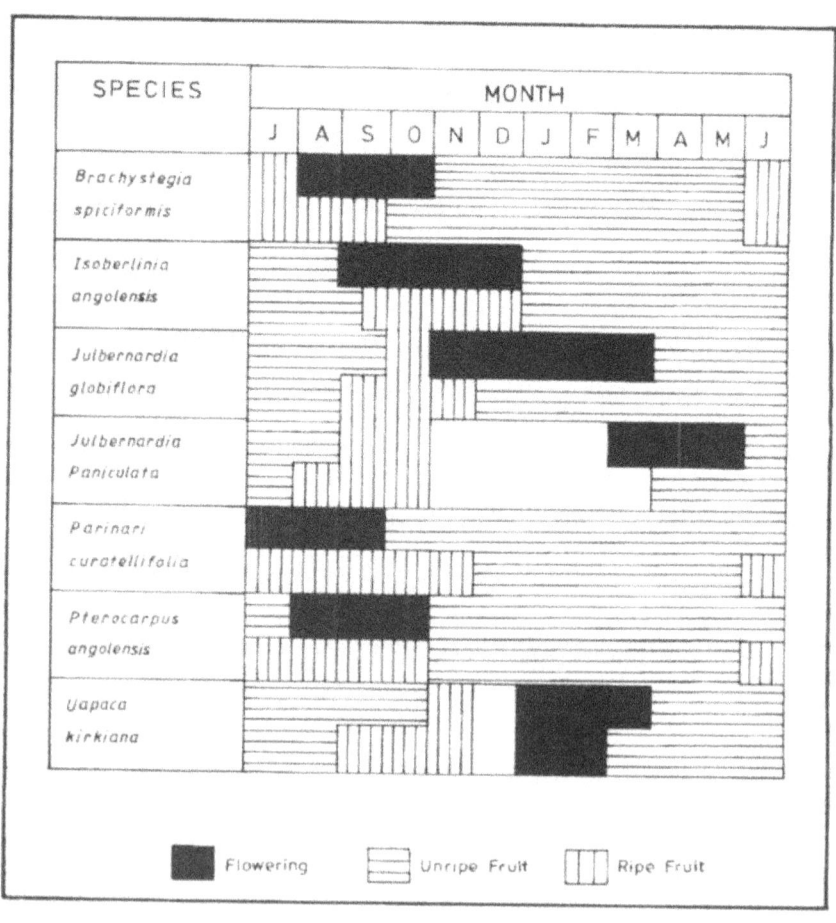

Figure 1.16 *Reproductive phenology of some dry miombo trees*

There is great diversity in fruit types and dispersal mechanisms among miombo trees (Table 1.12). Among canopy species the pod is the most common fruit type, and seed is either dispersed by an explosive pod (e.g., in *Brachystegia, Isoberlinia* and *Julbernardia* species) or by wind (e.g., in *Albizia* and *Pterocarpus* species). The fleshy fruit, including fruits with pulp, is common among understorey and shrub species, and animals, mainly birds and mammals, are the common dispersal agents. Malaisse (1978) found that the longest dispersal distances of 28–103 m were among wind-dispersed (anemochorous) species, followed by those with an explosive pod (autochorous species) with distances of 10–20 m, and the shortest distances (5–10 m) were among animal-dispersed (zoochorous) species.

Seed production and dispersal are not always good indicators of the seed crop available for germination, because of both pre-dispersal seed damage and predation. Very little is known about seed damage in miombo woodland. Chidumayo (1993a) found that the proportion of mature seeds that was damaged by either disease or predators before dispersal was 14 per cent in *U. kirkiana* and 17 per cent in *B. boehmii*. The highest seed loss due to predation was observed in *I. angolensis*: 75, 64, 70 and 86 per cent in 1990, 1991, 1992 and 1993, respectively. The main predators of *I. angolensis* seeds were bruchid larvae. Chidumayo (1993a) also observed that although pre-dispersal *J. globiflora* seed damage was 13 per cent, a survey of post-dispersal seeds revealed that 33 per cent were damaged by bruchid larvae. Consequently, the seed crop available for germination in miombo woodland is usually far less than actual seed production.

Seeds of the majority of miombo trees and shrubs germinate immediately after dispersal, as long as there is adequate water supply (Chidumayo 1991a, 1992b; Ernst 1988). Seed viability under uncontrolled storage conditions varies among miombo species (Table 1.13). *Brachystegia* and *Julbernardia* species lose viability within a few years, while no loss of viability was observed in *Afzelia quanzensis* and *Swartzia madagascariensis*. However, the storage of *B. spiciformis* seeds at 20 °C extended their viability to five years (Ernst 1988).

Table 1.13 *Seedling emergence rate (percentage per year) from seeds stored for varying periods under uncontrolled room conditions in five miombo woodland species*

	Storage period (years)			
	1	2	3	4
Afzelia qaunzensis	99	97	100	–
Brachystegia spiciformis	90	97	20	0
Julbernardia globiflora	73	35	0	0
Julbernardia paniculata	67	17	0	0
Swartzia madagascariensis	36	–	57	–

After Chidumayo (1993a)

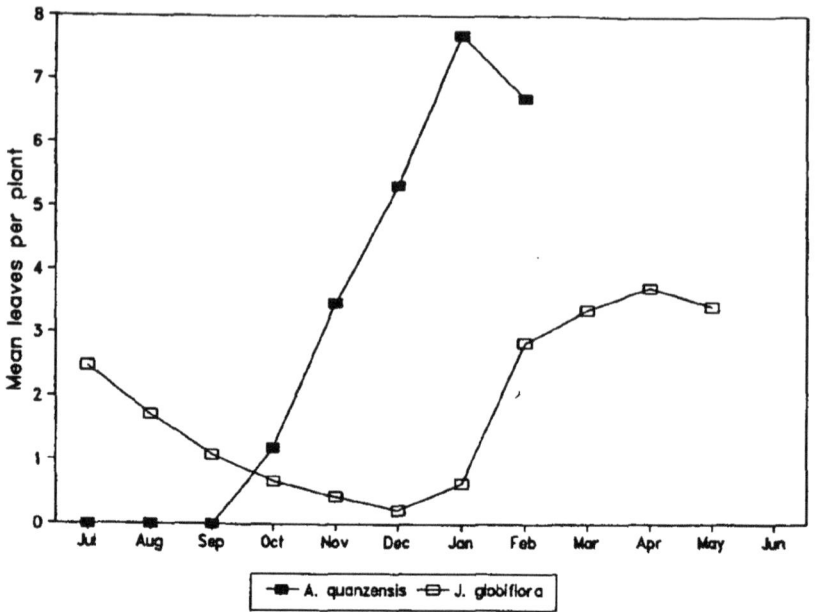

Figure 1.17 *Shoot growing season in seedlings of two miombo trees*

Seedling survival and development There is great inter-specific variation in survival rates of miombo tree seedlings, especially during the first year (Chidumayo 1991a, 1992b, 1992c). Under experimental field conditions, tree seedling mortality rates range from 12 per cent in *Afzelia quanzensis* to 22 per cent in *Brachystegia spiciformis*, 33 per cent in *Isoberlinia angolensis*, 55 per cent in *Julbernardia paniculata*, 67 per cent in *Julbernardia globiflora* and 100 per cent in *Uapaca kirkiana* during the first year. In subsequent years, the mortality rate declines to under 10 per cent (Chidumayo 1993a).

Tree seedling mortality in miombo has been attributed to drought, water stress and fire (Chidumayo 1991a, 1992b, 1992c; Ernst 1988; Strang 1966; Trapnell 1959). Different species resist these environmental stresses to varying degrees. Whereas *U. kirkiana* seedlings die from drought, *I. angolensis* seedlings survive both drought and fire (see Figure 4.13). Similarly, *B. spiciformis* seedlings lack fire resistance even in the second year, while the majority of *J. paniculata* seedlings are resistant to fire. In all cases, the shoots of the surviving seedlings are killed by fire but seedlings survive through roots which produce sucker shoots the following growing season.

The majority of seedlings of miombo trees experience a prolonged period of successive shoot die-back during their development phase. Shoot die-back is caused by water stress or fire during the dry season (Boaler 1966; Chidumayo 1991a, 1992b, 1992c; Trapnell 1959). All *J. globiflora* and *J. paniculata* seedlings die back during the first year compared to 67 per cent among *B. spiciformis* seedlings (Chidumayo 1991a, 1992b). However, shoot die-back does not necessarily result in seedling death if the root can survive and produce a new shoot the following growing season. This is an important adaptation in an environment that is subject to regular annual fires (Chapter 3) and drought.

Generally the seedlings of miombo trees grow slowly, even in the absence of shoot die-back (Chidumayo 1992b). During the prolonged seedling phase, shoot growth is largely confined to the rainy season, although the shoot extension period in saplings and trees occurs during the hot dry

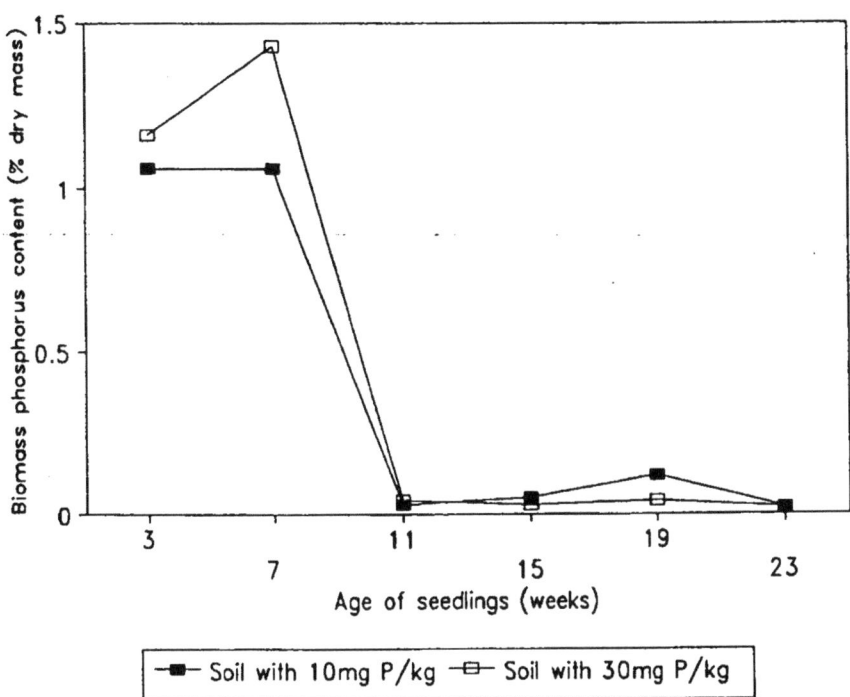

Figure 1.18 *Phosphorus content in Afzelia quanzensis seedlings grown in soil with differential concentration of available phosphorus*

Table 1.14 *Shoot growth during the establishment period of miombo woodland tree seedlings*

Species	Mean shoot height (cm)				
	1st year	2nd year	3rd year	4th year	5th year
Afzelia quanzensis	13.7	20.6	31.0	–	–
Brachystegia spiciformis	6.7	6.0	7.3	–	6.6
Julbernardia	5.3	4.9	6.8	9.3	–
Isoberlinia angolensis	9.2	6.7	–	–	–

Based on Chidumayo (1993a)

season (Boaler 1966; Chidumayo 1993a; Rutherford and Panagos 1982). *Afzelia quanzensis* is an exception, because the seedling shoot growing season is similar to that of saplings and adult trees (Figure 1.17). In general, however, seedlings in miombo grow slowly (Table 1.14). In contrast, *Acacia polyacantha* seedlings can attain a mean height of 43 cm and 186 cm by the end of the first and second years, respectively (Chidumayo unpublished). The slow shoot growth cannot be adequately explained by recurrent shoot die-back (Chidumayo 1992b) or nutrient deficiency, because seedlings do not appear to respond easily to mineral fertilizer application (Boaler 1966). In fact, the concentration of P, one of the limiting nutrients in miombo, in seedlings grown in soil with higher-available P was similar to those grown in normal miombo soil (Figure 1.18). These observations led Chidumayo (1992b) to suggest that slow shoot growth among seedlings of miombo trees is genetically based.

Roots of seedlings of many miombo trees grow faster than shoots during the establishment period (Chidumayo 1993a: Figure 1.19). Seedlings of miombo trees therefore allocate more photosynthetic biomass to root than shoot growth during the establishment phase. A comparison of ring counts of root stocks and their established shoots showed that at least eight years may be necessary for miombo woodland seedlings to reach the sapling phase (Lees 1962).

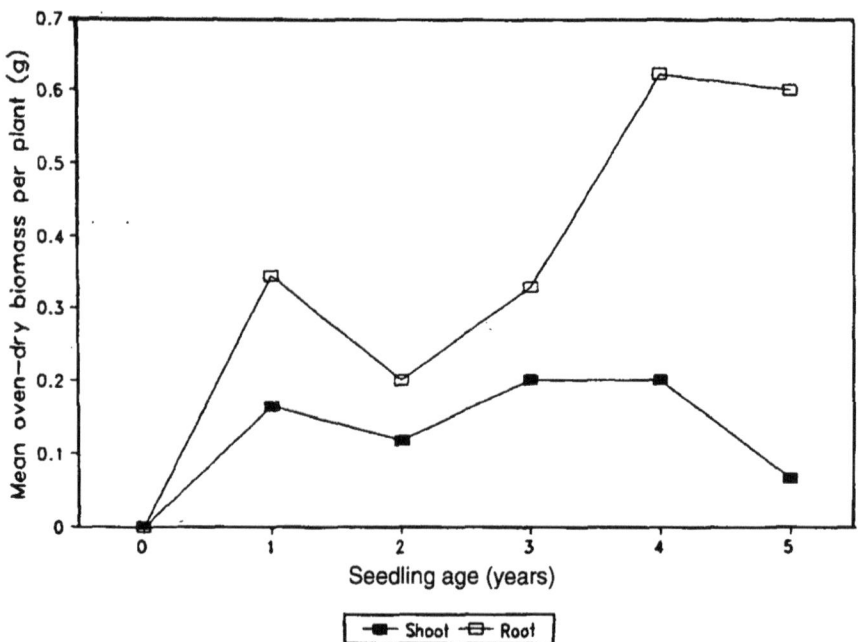

Figure 1.19 *Early development of shoot and root in seedlings of J. globiflora*

Table 1.15 *Changes in sucker shoots per plant during coppice establishment in dry miombo*

Species	Mean shoots per plant in regrowth of different ages (years)			
	1	3	9	18
Brachystegia boehmii	5.7	2.8	1.5	1.1
Julbernardia globiflora	3.9	1.8	1.1	1.1
Isoberlinia angolensis	6.6	2.5	1.3	1.0
All species	5.5	2.4	1.5	1.2

Based on Chidumayo (1993a)

Vegetative regeneration Stumps and roots of almost all miombo trees produce sucker shoots once the AG parts have been removed or killed (Banda 1988; Boaler and Sciwale 1966; Chidumayo 1989a; Hood 1972; Lees 1962; Strang 1974; Trapnell 1959). Sucker shoots arise from buds which develop on roots and stem bases. However, the capacity to coppice decreases with age and the size of the stem. In dry miombo, the survival survival rate of stumps in cut plots declined from 100 per cent in a 9-year old stand to 98 per cent in an 18-year old stand and 79 per cent in two old-growth stands (Chidumayo 1993a). In cut-over old-growth wet miombo plots, the stump survival rate ranged from 68 to 76 per cent (Chidumayo 1989a; Hood 1972).

Stumps produce many sucker shoots (Chidumayo 1988a, 1989a), but during the establishment period the number of shoots decreases due to intershoot competition, and only dominant shoots contribute to the next generation of regrowth miombo (Table 1.15). Stem density per plant therefore declines slowly with age of regrowth. However, fire may either slow or accelerate this domination process. If a destructive fire occurs before dominant shoots attain a safe height to escape mortality, the sucker shoot domination process is deflected back to the initial stage, and stumps respond by producing an equal or larger number of replacement shoots (Chidumayo 1988a).

Sucker shoots grow faster than shoots of stunted old seedlings. The average height of sucker shoots in dry miombo ranges from 35–65 cm, and exceptionally vigorous shoots may reach 150 cm height after one growing season (Chidumayo 1993a).

Regrowth Fanshawe (1971) noted that miombo regrows virtually unchanged following clearing. This is because regeneration consists of stump/root sucker shoots and recruitment from old stunted seedlings already present in the grass layer at the time of cutting. After one year the sapling population in regrowth may consist of one-third coppiced stumps and two-thirds seedlings recruited from the seedling pool (Table 1.16). As a result, tree density in regrowth miombo is higher. The high level of recruitment into the sapling phase from the seedling pool that has hitherto been stunted suggests that further development of seedlings with well-established roots is suppressed by the woodland canopy through shading. Lees (1962) also observed that old but stunted seedlings of many miombo woodland trees are heliophytic and require high light intensities to develop and grow. It follows therefore that succession in miombo is significantly affected by the 'initial floristic composition' factor (Egler 1952). An example of early succession at a dry miombo plot is given in Figure 1.20. The pre-felling stand and the seedling pool was dominated by *Julbernardia globiflora*, and this species continued to dominate the early succession in regrowth. This indicates that species that are dominant at the time of felling miombo are likely to dominate the regrowth. Little is known about late succession in miombo under natural conditions; however, fire (see Chapter 3) is known to alter the pattern of miombo regeneration and succession (Trapnell 1959).

Stem height increment in regrowth miombo is highest in the first or second year and declines thereafter (Table 1.17). The mean stem height may reach 4–5 m after 18 years in regrowth dry miombo (Figure 1.21). This subsequent slow height growth in regrowth miombo increases the risk of shoot mortality by wild fires. In addition, branches of many miombo trees tend to grow outwards and downwards after the stem has attained a certain height. Stem height is therefore a poor measure of shoot length increment in tall regrowth.

Table 1.16 *Woody plant (excluding suffrutices) density one year after cutting dry miombo. BD is basal diameter*

Type of woodland	Mean number of plants m^{-2}				
	Seedlings		Saplings (BD > 6mm)	Total	Live stumps
	Small (BD<2 mm)	Large (BD 3–5 mm)			
Old-growth	1.45	0.32	0.21	1.98	0.07
First regrowth	0.36	0.27	0.50	1.13	0.27
Second regrowth	0.12	0.27	0.77	1.16	0.53

After Chidumayo (1993b)

Table 1.17 *Stem height growth in regrowth dry miombo*

Species	Mean stem height (cm) in regrowth of different ages (year after cutting)					
	1	2	3	6	9	18
Brachystegia boehmii	39	62	100	251	254	329
Julbernardia globiflora	67	110	141	–	247	517
Isoberlinia angolensis	52	81	121	340	323	541
Uapaca kirkiana	60	75	–	245	268	460
All species	51	73	111	269	259	435

Based on Chidumayo (1993a)

The mean girth annual increment at 1.3 m above ground or breast height (gbh) in miombo trees in a 49-year old stand maintained under early and no burning regimes ranged from 1.1–1.5 cm and 0.6–1.0 cm for canopy and understorey species, respectively (Chidumayo 1988b). The corresponding figures for unmanaged 9-year old coppice plots were 1.4–2.2 cm and 1.1–1.7 cm per annum for canopy and understorey species, respectively (Chidumayo 1993a).

Table 1.18 *Above ground woody plant biomass in some Zambezian vegetation types in Zambia*

Vegetation type	Mean biomass (oven-dry t ha^{-1})				
	Leaf	Wood			Leaf plus wood
		Brush	Cord	Total	
Dry evergreen forest	7.0	29	158	187	194
Wet miombo	3.3	14	76	90	93
Central/eastern dry miombo	2.6	11	58	69	72
Western dry miombo	1.9	8	43	51	53
Mopane/Munga	1.9	7	38	45	47

Based on Chidumayo (1994a) and unpublished

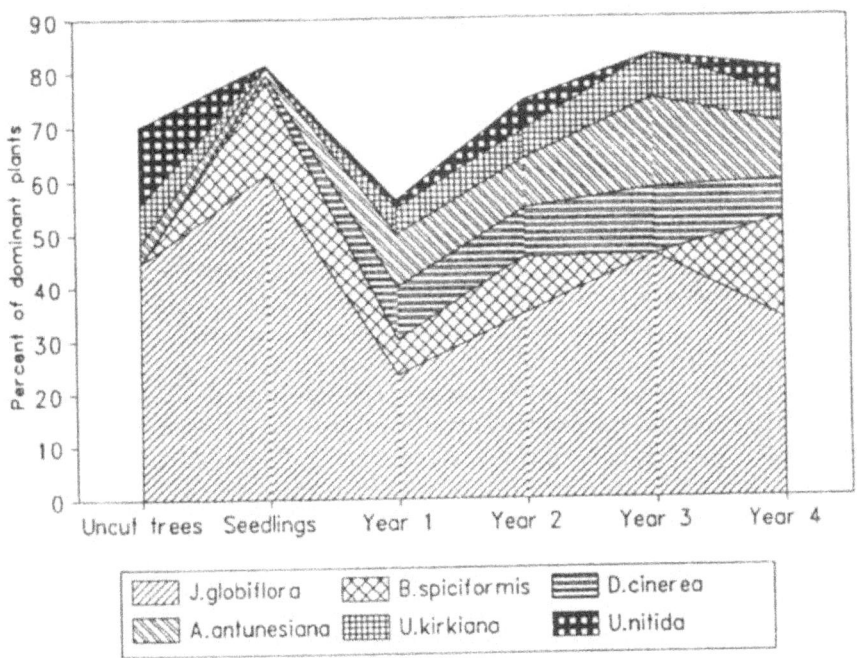

Figure 1.20 Early succession from uncut woodland to 4th year regrowth by dominant species after clearing a dry miombo plot in Zambia

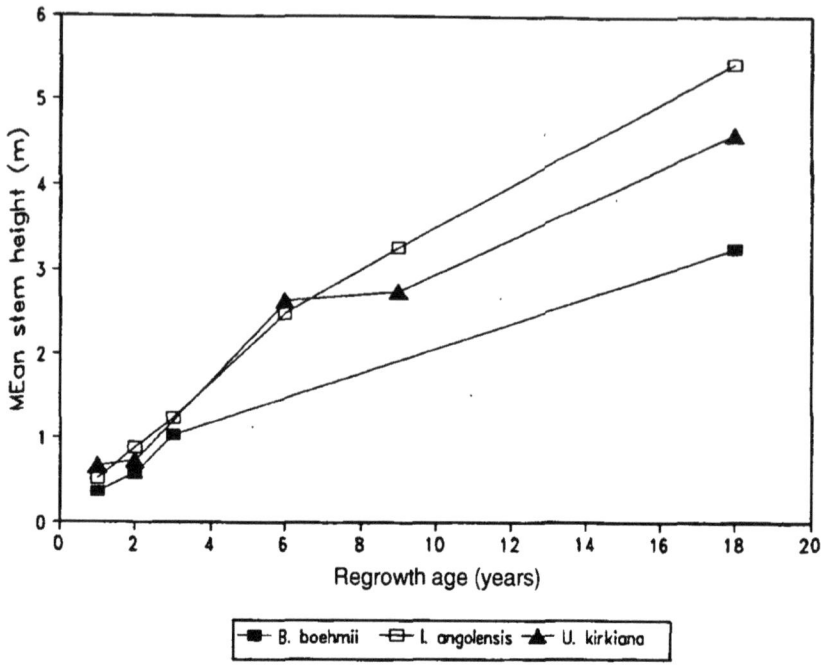

Figure 1.21 *Mean height growth of three miombo trees in regrowth dry miombo*

Woody plant biomass production

The growing season The tree growing season in miombo starts with leaf flush in the hot dry season (Table 3.4). Leaf flush is probably triggered by the rising mean temperature after the cool dry season, although other factors, such as increasing daily temperature range and day length, may also be involved (Rutherford and Panagos 1982). In the majority of miombo trees, leaf flush is preceded or accompanied by a drop in foliar N:P ratio (Figure 1.22) and the shoot extension period extends from September to December (Chidumayo 1993a; Rutherford and Panagos 1982) although production may continue until May or June. Tree growth is usually phased between shoots and roots. In seedlings of *Afzelia quanzensis* the shoot growing season ends in February while that of the root ends in April or May (Chidumayo 1992c). Similar phased growth has been observed in *Burkea africana* and *Ochna pulchra* trees in South Africa (Rutherford 1983). In these species the main period of root growth occurs towards the end of the rainy season and early in the cool dry season, and in *O. pulchra* cessation of root growth coincides with the onset of leaf fall, which probably also applies to the majority of miombo trees.

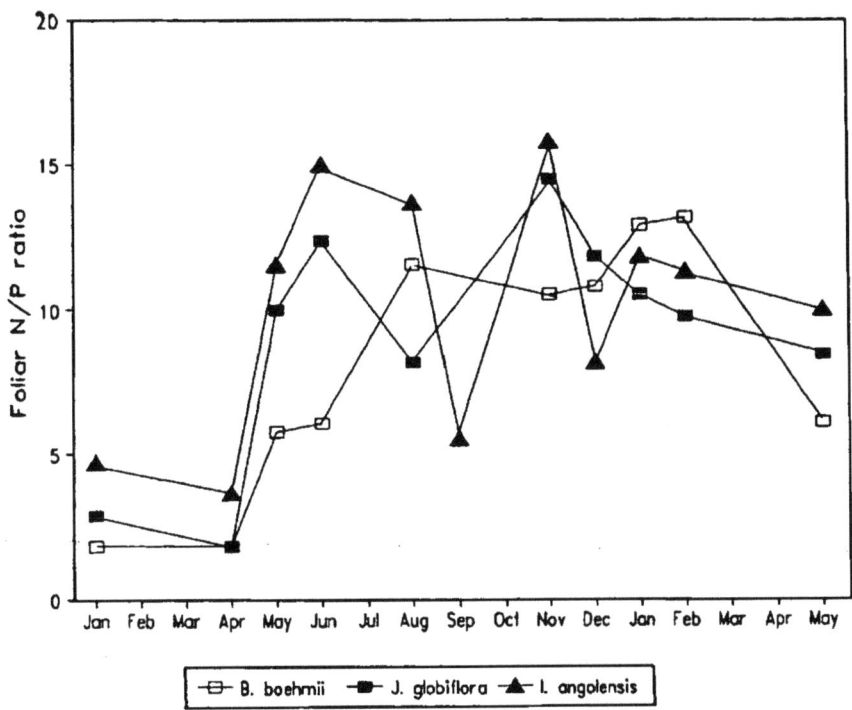

Figure 1.22 *Temporal changes in foliar nitrogen : phosphorus ratio in three miombo trees*

Biomass. Above ground (AG) woody plant biomass in old-growth uneven age miombo and other selected Zambezian vegetation types in Zambia is given in Table 1.18. Dry evergreen forest has the highest biomass, followed by miombo, while mopane and munga woodlands have the least biomass. Within miombo woodland, biomass increases from 53 t ha^{-1} in western dry miombo to 72 t ha^{-1} in central and eastern dry miombo and to 93 t ha^{-1} in wet miombo types. The low biomass in western dry miombo is probably related to lower nutrient status of the kalahari sand (Trapnell and Clothier 1957). The annual incremental growth in old-growth uneven age miombo has not been adequately studied, but in central dry miombo this has been estimated at 1 per cent of the AG wood standing biomass (Chidumayo 1993a). The only available below ground (BG) woody plant biomass estimate is that of 20 t ha^{-1} in central dry miombo which represents about 60 per cent of AG biomass (Chidumayo 1993a).

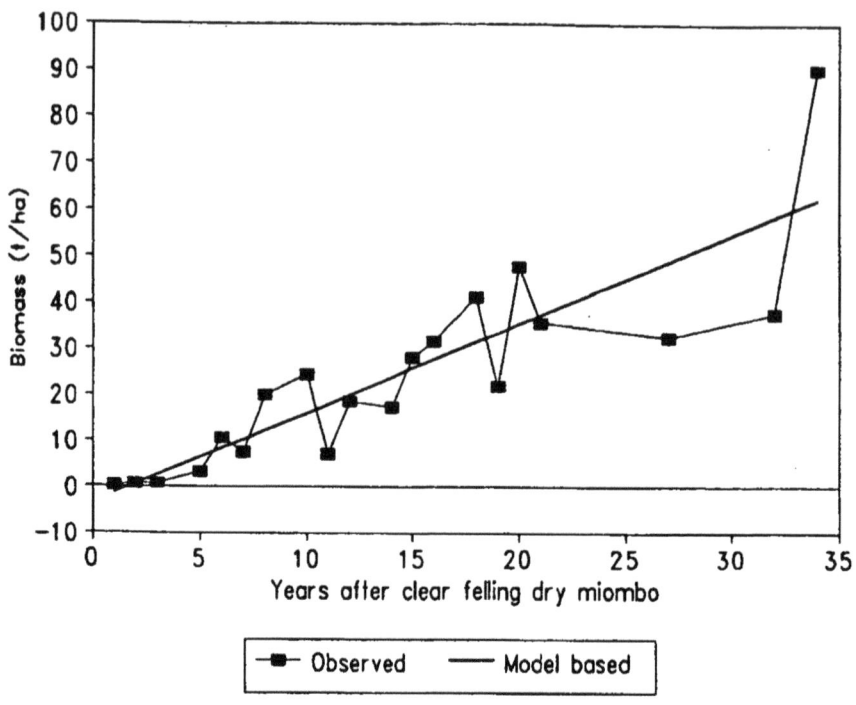

Figure 1.23 *Correlation between above ground wood biomass and age of coppiced dry miombo*

There is a positive correlation between age and leaf production and wood annual increment in regrowth miombo during the first 10 years. However, this correlation disappears in older regrowth (Figures 1.25 and 1.26). Production of leaf and wood biomass is similar during the initial ten years but trees allocate more to leaf production in older regrowth (Figures 1.25 and 1.26). In both dry and wet regrowth miombo the regression of AG wood biomass on age is significant (r-squared is 0.74: Figures 1.23 and 1.24). However, the average wood annual increment of 2.7 t ha^{-1} yr^{-1} in wet miombo is higher than that of 1.9 t ha^{-1} yr^{-1}. Very little is known about BG biomass production, although a limited study in central dry miombo estimated a mean annual increment of 0.6 t ha^{-1} yr^{-1} (Chidumayo 1993a).

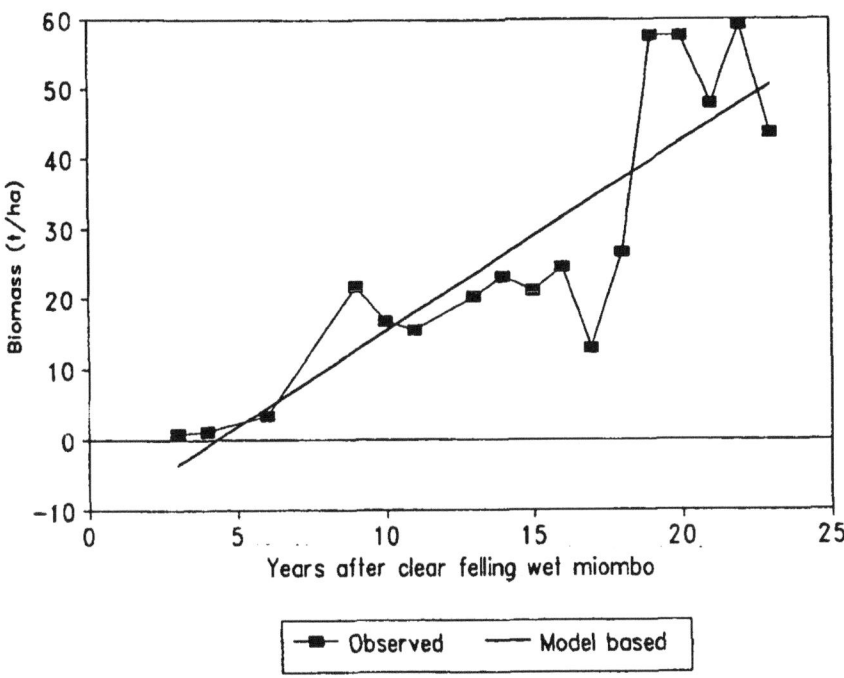

Figure 1.24 *Correlation between above ground wood biomass and age of coppiced wet miombo*

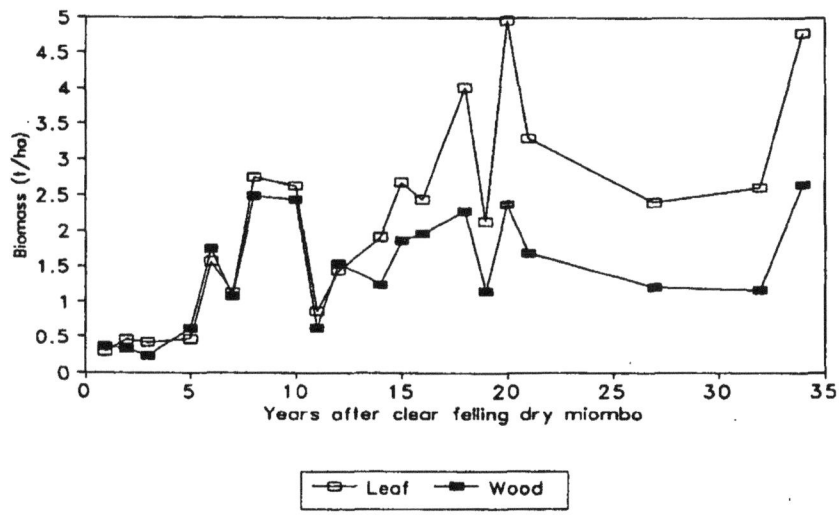

Figure 1.25 *Age-related changes in leaf production and mean wood annual increment in coppiced dry miombo*

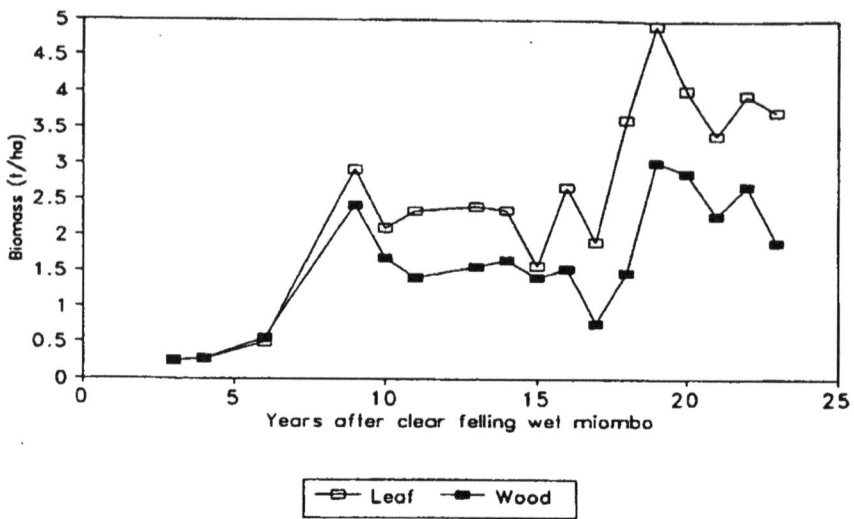

Figure 1.26 *Age-related changes in leaf production and mean wood annual increment in coppiced wet miombo*

CHAPTER 2
The Utilization of Miombo

MODERN MAN probably inhabited present-day Zambia from the Lower Pleistocene period, about 2 million years ago, but evidence of human settlements in the miombo region only dates back some 10 000 years ago. The pattern of ecosystem use intensified with each cultural succession, starting with the Late Stone Age, which had the least impact, through the Iron Age to the present Industrial Age. Human activities increased with the emergence of the Early Iron Age culture in the Late Holocene, when agriculture and permanent and semi-permanent settlements evolved. The Iron Age was also associated with large-scale population movements and witnessed the arrival of the Negroid people into Zambia around 2-4 AD. The migrations of the Luba and Lunda peoples from Zaire during 1500–1700, and those of the Ngoni from South Africa in the nineteenth century (Langworthy 1971), further increased the use of the miombo ecosystem, as did the rise of urbanization and industrialization during the colonial era. With an annual population growth rate of 3.2 per cent during 1980–1990, human pressure on natural ecosystems in Zambia has continued. Table 2.1 shows the current distribution of the Zambian population.

Table 2.1 *Distribution of human population in Zambia in 1990 by residence*

Province	Area (sq.km)	Population (*1000)		Households		Rural population density
		Urban	Rural	Urban	Rural	(no/sq.km)
Central	94 390	216	510	36 000	96 226	5.40
Copperbelt	31 320	1 429	151	238 167	28 491	4.82
Eastern	69 100	86	888	14 333	167 547	12.85
Luapula	50 560	83	536	13 833	101 132	10.60
Lusaka	21 890	1 041	167	173 500	31 509	7.63
Northern	147 820	123	744	20 500	140 377	5.03
N-western	125 820	46	338	7 667	63 774	2.69
Southern	85 280	190	756	31 667	142 642	8.86
Western	126 380	71	536	11 833	101 132	4.24
Zambia Total	752 560	3 285	4 626	547 500	872 830	6.15

Based on various report by the Central Statistics Office

Agriculture

Cultivation

Although dry miombo soils are relatively fertile and therefore have a higher agricultural potential than wet miombo soils (Tables 1.6 and 1.7), in general these soils are inherently infertile. In fact, soils in Zambia are not well suited for permanent cultivation. For example, although 42 million ha are potentially arable (Bunyolo *et al.* 1993), only 10 per cent of the country (7.5 million ha) is suitable for maize production, the staple food for urban areas and Central, Eastern and Southern Provinces (Appendix 1). Because of this environmental constraint, traditional cultivation systems rely on a natural fallow being allowed after a period of cultivation, as a mechanism for regenerating soil fertility. The soils in the high rainfall region are also more acid, with aluminum toxicity, which further constrains crop production.

The different traditional cultivation systems in Zambia have been described by Trapnell (1953), Trapnell and Clothier (1957) and Schultz (1974). Basically these fall into two main classes: shifting and semi-permanent. The shifting cultivation system is subdivided into extensive and intensive forms. The extensive system is further subdivided into large- and small-circle chitemene (chitemene means 'to cut').

The chitemene cultivation system is predominantly practised in northern wet miombo. Crops are grown, without tilling the soil, in an ash garden (infield) made from the burning of a pile of branches obtained by lopping and chopping trees from an area (outfield) 8–10 times larger than the ash garden (Chidumayo 1987b; Stromgaard 1989). By the lopping technique only 30 per cent of the AG woody biomass is harvested for ash fertilization (Table 2.2), thereby allowing the stumps and trunks in the outfield to immediately initiate the process of regeneration back to woodland. The pile of woody biomass is burnt in October, just before the onset of the rainy season to translocate the nutrients from the biomass to the ash garden. In spite of losses during burning, the ash contains considerable amounts of nutrients. For example, Stromgaard (1984) found that the ash on a chitemene infield contained 44, 1 and 219 kg ha^{-1} of N, P, and K, respectively. In fact there is a positive correlation between ash and mass yield of finger millet (*Eleusine coracana*), the primary crop of new ash gardens in northern Zambia (Araki 1992). In addition the ash increases soil pH while the sterilizing heat generated during biomass burning reduces soil microbes that would normally effectively compete with crops for nutrients (Chidumayo 1987b). Araki (1992) has shown that the effect of additional exchangeable bases (Ca, K, Mg, and Na) from the biomass on millet yield is twice as large as that of heat.

Table 2.2 *Above-ground woody plant biomass portioning for chitemene cultivation in northern Zambia*

Biomass component	Oven-dry biomass	
	t ha^{-1}	% of total
Stumps	18.84	20.43
Trunks	44.74	48.50
Branches	26.23	28.44
Leaves	2.43	2.63
Total	92.24	100.00

After Araki (1992)

Traditionally a new ash garden is made every year (Richards 1939; Allan 1949; Trapnell 1953). In the second year cassava, which matures over a period of 2–3 years, succeeds millet before the plot is abandoned. The abandonment is triggered by soil reacidification to the pre-burn level (Lungu and Chinene 1993). The sustainable population density in the large-circle chitemene is estimated at 2–4 km^{-2} (Chidumayo 1987b: see Chapter 4 for the estimation of critical population density under chitemene cultivation). This density has obviously been exceeded in both Luapula and Northern Provinces (Table 2.1), where large-circle chitemene is practised. Due to this population pressure and a shortage of old-growth woodland, households now clear woodland for chitemene once in two years (Stromgaard 1985a; Table 2.3). In addition, fallow periods have been reduced from about 25 years to 12 years (Chidumayo 1987b). As a consequence, trees are now cut at breast or knee height instead of overhead lopping. However, average woodland clearings have remained at 8.3 ha for an average ash garden (Stromgaard 1989). Since new clearings are done once in two years, the annual woodland requirement for a household has declined to 4.15 ha.

Table 2.3 *Trends in woodland conversion to chitemene in a 250-ha block of miombo in Bwacha village in Kasama district, Northern Province, Zambia*

Year	Woodland clearing for chitemene cultivation	
	Cut-over area (ha)	Cultivated area (ha)
1975	1.88	0.28
1976	4.00	1.00
1977	19.58	2.60
1978	2.65	0.40
1979	21.35	3.00
1980	0.45	0.05
1981	27.25	30.50
Total	77.16	10.83
% of woodland	30.86	4.33

Based on Stromgaard (1985a)

In north-east Zambia fallow periods are so short that grass and young coppice has replaced woodland as a natural fallow (Stromgaard 1989). Most of the grass is cut and composted in mounds (fundikila system), while excess grass is heaped around tree stumps and later burnt. This system, which permits higher population densities (12–20 km^{-2}), depends on the transfer of nutrients in the grass biomass (Table 2.4) through decomposition by microbes. Soil fertility is maintained by N-fixation by symbiotic bacteria in the roots of legume crops which are grown in rotation with cereal crops (maize and millet). A legume crop (beans or groundnuts) is sown on the mounds, and in the following season the mounds are broken and the mound soil spread to form a flat bed on which millet and other cereals are grown. This alternation of mound and flat cultivation is carried out for 4–6 years without any significant change in the soil nutrient pool (Table 2.4).

Table 2.4 *Changes in top-soil (0–10 cm depth) nutrient pool during the fundikila grass-mound cultivation cycle in north-eastern Zambia*

	Nutrient						
	Org. C (%)	Total N (%)	Mean P (ppm)	Exchangeable bases (meq/100 g)			
Cultivation stage				Ca	Mg	K	Na
Fallow grassland	1.00	0.04	2	1.20	1.40	0.20	0.04
Mounds (2-month old)	0.90	0.04	4	1.20	1.00	0.30	0.02
Mounds (1-year old)	1.07	0.04	1	3.01	0.30	0.12	0.05
Flat (2-year old)	0.70	0.04	2	1.10	0.80	0.10	0.02
Flat (4-year old)	1.10	–	4	2.40	1.10	0.20	0.02

After Stromgaard (1990)

The introduction of hybrid maize production in Northern Province in the 1980s, based on cheap subsidized inorganic acid fertilizer without lime, had some negative impacts on the ecosystem. The maize production system encouraged tree clearing at ground or underground (uprooting) level and soil tillage either by hand hoes or ox-drawn and tractor equipment. The ploughing brings the more acid subsoil onto the surface and increases the risk of rapid acidification from the use of acid fertilizer. Furthermore, although the initial maize production is high, it declines to uneconomic levels within four years (Lungu and Chinene 1993: Figure 2.1), thereby necessitating fallowing almost as frequently as under the traditional chitemene system. However, because of the progressive lowering of the tree-cutting height from overhead lopping, to breast or knee height, to ground or underground cutting, the woodland recovery period is progressively increased and the pattern of regeneration changed, with root suckers assuming a greater role in the regeneration process.

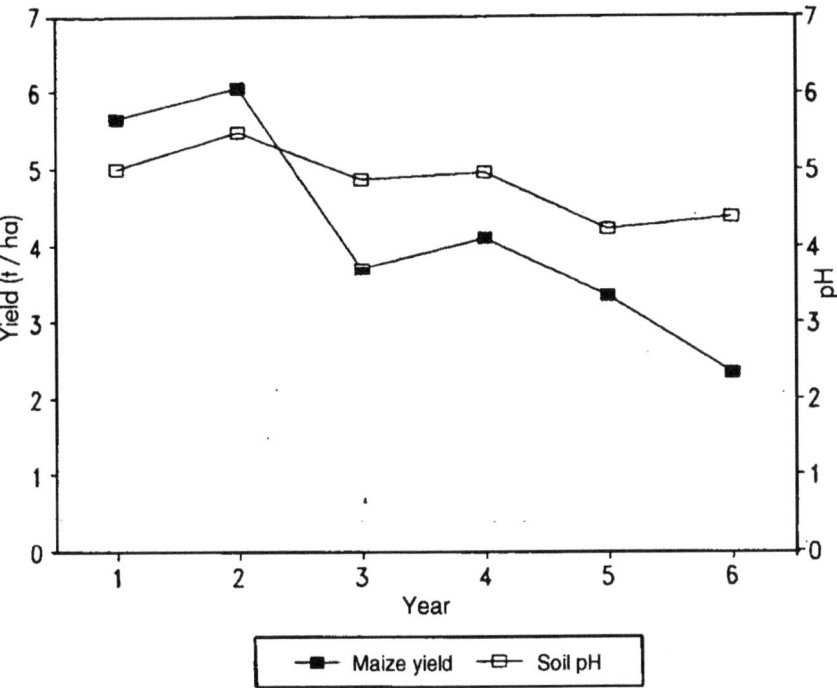

Figure 2.1 *Decline in maize yield and soil pH under monocropping with continuous application of nitrogenous fertilizer at Misamfu in northern Zambia*

Small-circle chitemene is practised in the southern parts of the chitemene region and in particular the northern districts of Central Province. This differs from the large-circle chitemene in two main ways. First, the branches are cut at breast height and piled in several small circles or narrow long strips, and cultivation is by hoe. Second, the ash garden is only 4 per cent of the cleared land compared to about 10 per cent under the large circle, and is cultivated for only 2 years before abandonment (Stromgaard 1989). The small-circle chitemene system's impact on the ecosystem is therefore similar to a large-circle chitemene under population pressure, with shorter fallow periods when trees are cut at breast height.

The intensive forms of traditional shifting cultivation are found in the Copperbelt, North-western and Western Provinces and the escarpment and valley areas of Central, Eastern and Southern Provinces. Trees are normally cut at breast height and either burnt *in situ* or in piles, but the whole area is cultivated by hoe. Sometimes annual extensions are made by clearing adjacent woodland, but more often new areas are cleared only when the old cultivations are exhausted. This occurs after 3–4 years (Stromgaard 1989).

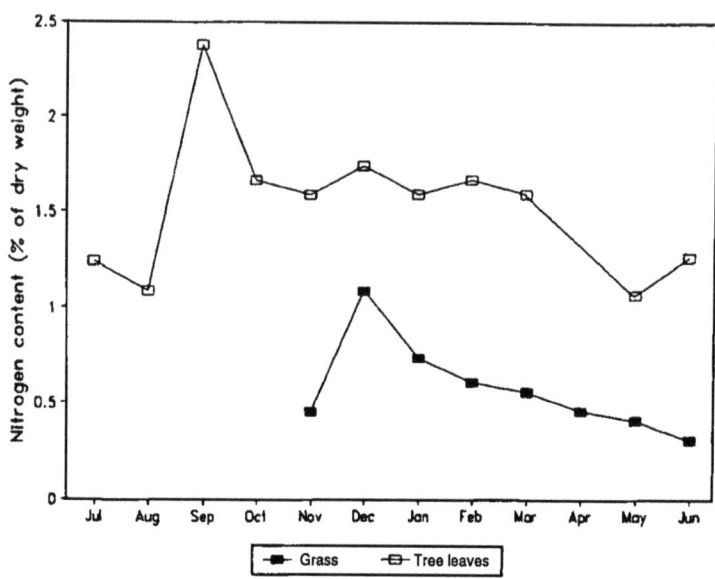

Figure 2.2 *Monthly changes in nitrogen content in grass and tree leaves in dry miombo*

The semi-permanent cultivation system is a further development of the intensive shifting system and is practised in plateau dry miombo of Central, Eastern and Southern Provinces. The colonial government encouraged this type of agriculture to achieve semi-commercialization of traditional cultivation in areas with higher agricultural potential to feed the growing urban population in the Copperbelt Province. Basically, crop production is maintained by supplying nutrients from inorganic fertilizer or animal manure. Since soil preparation involves ox-drawn and/or tractor implements, trees are normally uprooted and the whole area cultivated for 5–10 years before fallowing. This extended cultivation period increases the woodland recovery period. Very short fallow periods (under 10 years) are used, and soil erosion and watershed degradation (especially in conjunction with pastoralism) have become important environmental problems in this farming system. However, the higher levels of crop production have enabled greater population densities (10–20 km^{-2}) to be sustained.

Marter and Honeybone (1976) found that the average cultivated area in traditional shifting cultivation systems was 1.28 ha per household, compared to 4.14 ha in semi-permanent cultivation. However, the area under cultivation is usually less than the actual farm holding. In the chitemene region, the cultivated plot is 4–10 per cent of the deforested area (Stromgaard 1989), and in semi-permanent cultivation only 43 per cent of the farm holding is actually cultivated in any one year (Kwesiga and

Table 2.5 *Cultivated areas by vegetation and traditional farming system in Zambia in 1990*

Vegetation type	Potential extent of cultivation km²		Shifting cultivation km²		SPC¹ km²	Total cultivated area	
	km²	%	Chitemene	Other		km²	%
Dry forest/Kalahari miombo	128 500	17.4	0	1 520	0	1 520	6.4
Other forests	3 200	0.4	0	0	0	0	0.0
Miombo	308 600	41.7	2410	1 585	12 984	16 979	71.6
Mopane/Munga	92 200	12.5	0	1 033	2 662	3 695	15.6
Grassland	206 800	28.0	0	1 514	0	1 514	6.4
Total	739 300	100.0	2410	5 652	15 646	23 708	100.0

Based on data from various sources
1:Semi-permanent cultivation

Chisumpa 1992). The use of the area under cultivation therefore underestimates deforestation caused by cultivation in Zambia (Table 2.5). When fallow areas are included, estimates put the area under cultivation at 28 per cent in the chitemene farming region, 12 per cent in the intensive shifting farming region and 40 per cent in the semi-permanent farming region (Schultz 1974). A recent survey of 6.84 million ha of the plateau region in Central, Copperbelt and Lusaka Provinces put the cultivated and fallow land at 13 per cent (World Bank 1990). Conversion of miombo to cultivation is therefore a significant cause of deforestation in Zambia, and 72 per cent of the cultivated land is in miombo woodland (Table 2.5).

Pastoralism
The co-existence of woody and herbaceous plants in miombo enhances the potential for pastoralism. Grass production in miombo is confined to the rainy season (see Figure 4.9) and the quality of grazing declines as the growing season progresses. In contrast, woody plants flush 2–3 months before the onset of the rainy season and the young foliage constitutes a significant source of nutritious fodder for livestock (Campbell et al. 1991; Hood 1972). This differential phenology of woody and herbaceous plants ensures the availability of quality fodder almost throughout the year (Figure 2.2).

Important livestock areas in Zambia are Central, Eastern, Southern and Western Provinces. The Western Province, with the largest cattle population, has witnessed a steady increase in bovine population since 1964 (Figure 2.3). It is now feared that future growth will be limited by inadequate grazing in the woodland (van Gils 1988). The development of management techniques to promote browsing may therefore play an important role in sustaining cattle population growth. Hood (1972) identified a total

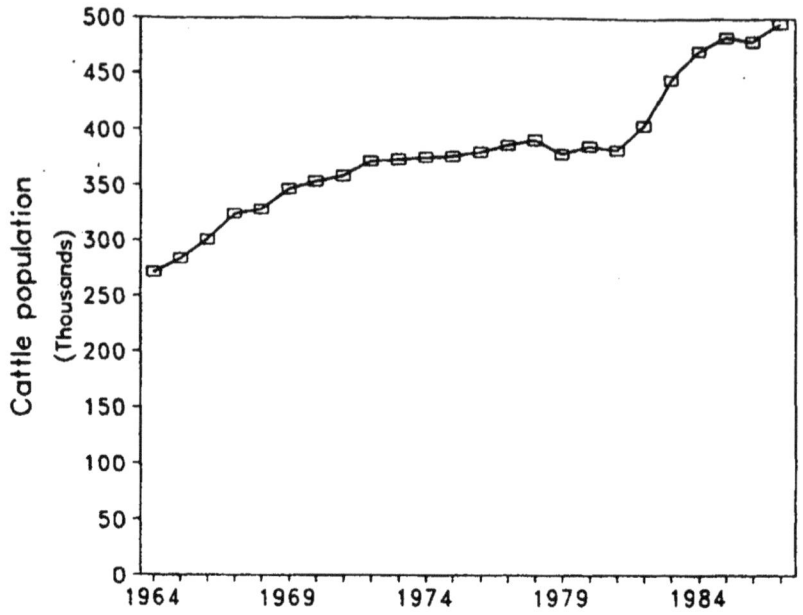

Figure 2.3 Growth in cattle population during 1964–87 in Western Province, Zambia

Table 2.6 Average exploitable timber stocking rate in north-western wet miombo in the Zambian Copperbelt

Stem girth (1.3 m AG) size class	Average exploitable stock ha^{-1}	
	Volume over bark (m^3)	Mass (t)
45–107cm	3.75	2.34
>107 cm	2.13	1.33
Total	5.88	3.67

Based on Lees (1992)

of 14 browse species in a 1.24-ha wet miombo stand, but the eight most palatable species were *Baphia bequaertii, Brachystegia spiciformis, Parinari curatellifolia, Isoberlinia angolensis, Julbernardia paniculata, Brachystegia floribunda, Diplorhynchus condylocarpon* and *Uapaca nitida. Albizia antunesiana, Julbernardia globiflora, Burkea africana, Dichrostachys cinerea, Pericopsis angolensis, Swartzia madagascariensis, Piliostigma thonningi* and *Ziziphus abyssinica* are also important fodder species in dry miombo.

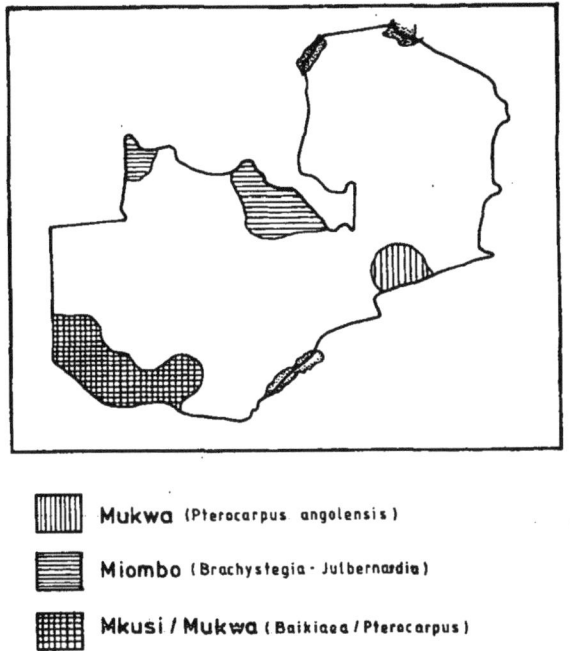

Figure 2.4 *Main commercial timber areas in Zambian indigenous forests*

Wood products

Of the five main commercial timber areas in Zambia, two are in wet rniombo in the Copperbelt and Northwestern Provinces, one is in *Baikiaea* dry deciduous forest in Western Province and another in munga woodland in Petauke district of Eastern Province (Figure 2.4). The dry deciduous forest has been selectively exploited for teak (B. *plurijuga*) and *Pterocarpus angolensis* since the 1920s but logging declined considerably since the 1970s (Figure 2.5). Wet miombo in the Copperbelt Province was also selectively exploited for P. *angolensis* and species of *Albizia, Faurea, Brachystegia* and *Julbernardia* for sawlogs and mine poles (Figure 2.6). The timber stocking rate in the Copperbelt Province assessed by Lees (1962) was estimated at 3.7 t ha^{-1} (Table 2.6). This represents about 4% of the AG wood biomass of 90 t ha^{-1} (Table 2.2). In dry miombo average wood biomass is 70 t ha^{-1} and the timber and pole mass has been estimated at 4.7 t ha^{-1} (Edmonds 1964). Timber therefore represents about 7 per cent of the standing wood biomass in dry miombo. In many parts of the miombo region poles of variable sizes are used for building houses, huts and other structures with clear species preferences (Table 2.7). The size and abundance of the poles depends on the age of the woodland (Table 2.8).

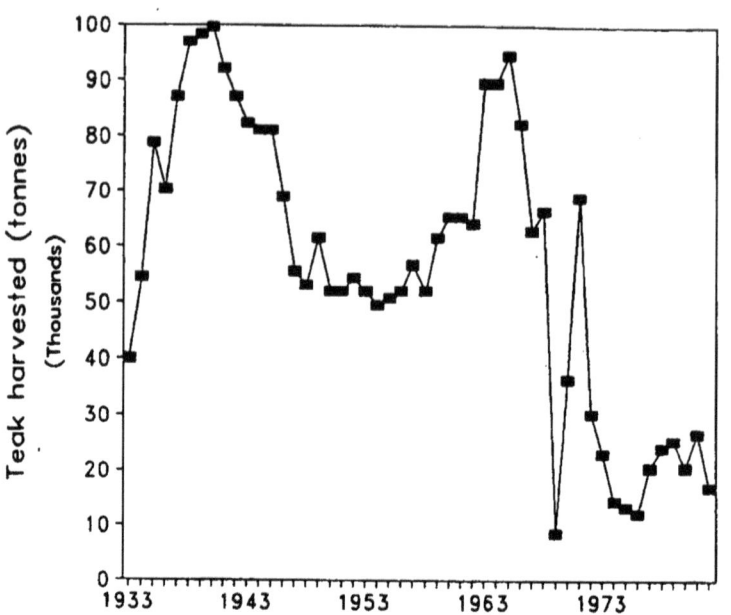

Figure 2.5 *Teak (*Baikaea plurijuga*) harvesting in southwestern Zambia 1933–82*

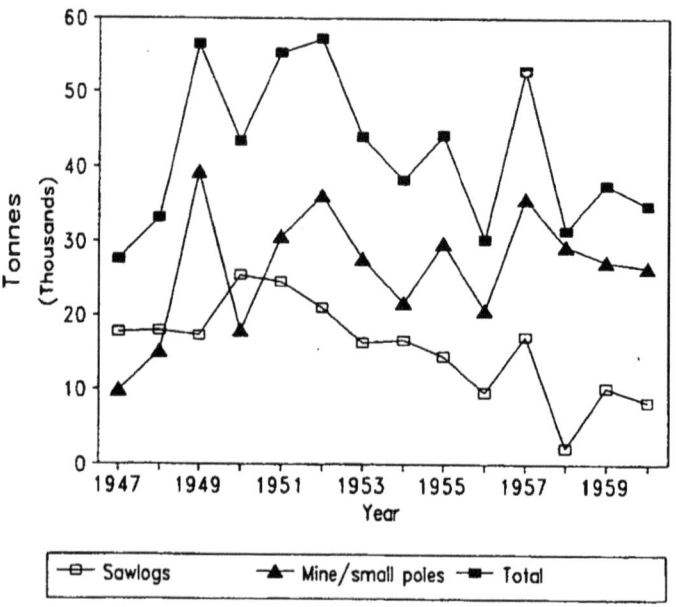

Figure 2.6 *Timber harvesting in wet miombo in the Zambian Copperbelt 1947–60*

Table 2.7 *Tree species preferred for building and other purposes in miombo woodland by the Bemba of northern Zambia*

Purpose	Tree species
Fibre	*Brachystegia boehmii*
	Brachystegia longifolia
	Brachystegia taxifolia
	Julbernardia globiflora
Handles	*Burkea africana*
	Dalbergia nitidula
	Isoberlinia angolensis
	Julbernardia globiflora
	Julbernardia paniculata
	Lonchocarpus capassa
	Swartzia madagascariensis
Mortar	*Albizia antunesiana*
	Isoberlinia ango/ensis
Poles	*Erythrophleum africanum*
	Monotes sp.
	Pericopsis angolensis
	Swartzia madagascariensis
Walling	*Brachystegia longifolia*
	Julbernardia globiflora
	Julbernardia paniculata
	Monotes sp.
	Uapaca kirkiana
	Uapaca nitida
Roofing	*Diplorynchus condylocarpon*
	Marquesia macroura
	Monotes sp.
	Pericopsis angolensis
	Syzygium owariense
	Uapaca kirkiana
	Uapaca nitida

Based on Holden (1988)

Table 2.8 *Average length (m) and density (no. ha^{-1} of poles in central dry miombo in Zambia*

	Whole stem poles in regrowth		Uneven-age	Old-growth
Variable	6–10 yrs	16–20 yrs	Whole stem poles	Bole poles (trunk) only
	(3 plots)	(3 plots)	(16 plots)	(16 plots)
Mean length (m)	1.33	2.61	4.98	2.81
SD	0.54	1.38	2.46	0.35
Mean density	3125	2075	690	42
SD	815	479	191	11

Based on Chidumayo (unpublished)

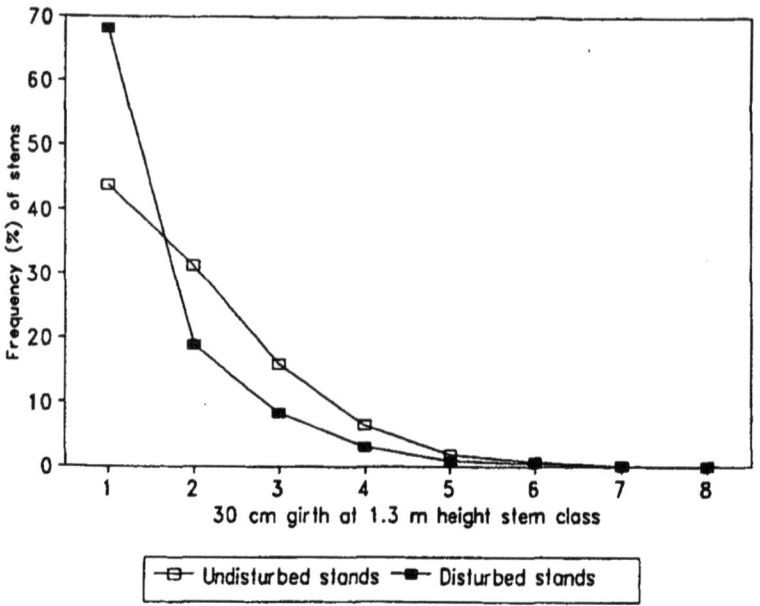

Figure 2.7 *Changes in stem size structure in dry miombo caused by selective felling*

Selective harvesting tends to alter woodland structure (Chidumayo 1987c) and reduces the genetic diversity of timber species because the best trees with straight poles are selected for. This disturbance is characterized by a high predominance of small stems in selectively felled areas (Figure 2.7).

There is a growing interest in Zambia in promoting commercial wood carvings and other hand crafts. The most extensively used species for this purpose are *Combretum zehyeri* (roots used for basket making), *Pterocarpus angolensis* (wood for carvings and musical instruments) and *Hyphaene petersiana* palm (leaves for basket making). Little is known about the impact of these activities on the species. However, in the Kalabo district of Western Province, *C. zehyeri* roots are utilized for making baskets. This small tree has long lateral roots which may reach 20 m. Although the roots of one tree are harvested once per year, they are cut too close to the crown and the furrows are not filled in after harvesting (Kotze 1993). Regrowth roots become exposed which may negatively affect the recovery of the harvested trees.

Woodfuel

The Zambia national energy budget is dominated by biomass which accounts for 76 per cent and 69 per cent of the primary energy supply and final consumption, respectively (Table 2.9). Biomass energy is largely used as a household energy source and primarily for cooking. According to the Department of Energy (1992), the household sector used 89 per cent of the biomass fuels totals (firewood and charcoal) during 1986–1990 while 4 per cent and 7 per cent were used in agriculture and industry. In rural areas the primary energy source is firewood and rural households and agriculture use 79 per cent and 6 per cent, respectively, of this fuel while urban households and industry use 6 per cent and 9 per cent. Charcoal is predominantly an urban fuel and urban households and industry consume 85 per cent and 45 per cent, respectively, of this fuel while rural households use 11 pre cent. No charcoal is used in agriculture (Department of Energy 1992). The average annual consumption of firewood is estimated at 5,000 kg and 635 kg per household in rural and urban areas, respectively.

The consumption of charcoal averages at 100 kg per rural household per year and 1,040 kg per urban household. Based on data in Table 2.1, estimates of woodfuel consumption in rural and urban Zambia in 1990 are given in Table 2.10. Rural subsistence firewood collection rarely affects the miombo woodland structure because only dead wood or wood cut for other purposes is collected. The collection of dead wood reduces the amount of wood debris in the woodland. Shackleton (1993) observed that woodland subjected to rural firewood collection had a dead wood biomass of 128 kg ha^{-1} compared to 745 kg ha^{-1} in a protected area in a South African savanna.

Table 2.9 *National energy supply and consumption (in tonne oil equivalent) in Zambia in 1990*

Energy source	Primary supply		Final consumption	
	Quantity	%	Quantity	%
Firewood	2 500 000	43	2 500 000	58
Charcoal	1 900 000	33	480 000	11
Electricity	580 000	10	520 000	12
Coal	260 000	4	260 000	6
Petroleum	570 000	10	570 000	13
Total	5 810 000	100	4 330 000	100

Department of Energy (1992)

Table 2.10 *Household woodfuel consumption (× 1000 tonnes) in Zambia in 1900, by residence*

	Rural areas				Urban areas			
	Charcoal	Wood			Charcoal	Wood		
Province		Firewood	Charcoal	Total		Firewood	Charcoal	Total
Central	9.6	481.1	41.8	522.9	36.0	22.9	156.5	179.4
Copperbelt	2.9	142.5	12.4	154.9	238.2	151.2	1,035.5	1,186.7
Eastern	16.8	837.7	72.9	910.6	14.3	9.1	62.3	71.4
Luapula	10.1	505.7	44.0	549.6	13.8	8.8	60.2	68.9
Lusaka	3.2	157.6	13.7	171.3	173.5	110.2	754.4	864.5
Northern	14.0	701.9	61.0	762.9	20.5	13.0	89.1	102.2
N-western	6.4	318.9	27.7	346.6	7.7	4.9	33.3	38.2
Southern	14.3	713.2	62.0	775.2	31.7	20.1	137.7	157.8
Western	10.1	505.7	44.0	549.6	11.8	7.5	51.5	60.0
Zambia	87.3	4,364.2	379.5	4,743.6	54.8	347.7	2,380.4	2,728.1

Based on data from various sources

The cutting of live trees for fuel in miombo has been associated with the emergence and growth of urban woodfuel markets. Urban towns in Zambia emerged during the 1930s, following the discovery of large copper deposits in the Copperbelt earlier in the twentieth century. In the past, the copper industry was dependent on miombo for woodfuel for electricity generation and charcoal in the refineries. During 1947–1956 about 127 000 ha were clear-cut (completely cleared of timber) for electricity generation (Lees 1962) and the copper industry still uses about 2000 t of charcoal annually in its refineries.

Since its introduction in the Copperbelt in 1947, charcoal has become a major urban household fuel in urban Zambia. Miombo woodland is the main source of wood used in charcoal production (Chidumayo and Chidumayo 1984) and almost all the charcoal in Zambia is produced by the traditional earth-kiln method (Chidumayo 1991b; Chidumayo and Chidumayo 1984; Ranta and Makunka 1986; World Bank 1990). The method involves tree felling, stem cross-cutting, kiln building and covering, wood carbonization, kiln tending and breaking to recover the charcoal. The charcoal is produced by individuals on either a part- or full-time basis. Production is typically on a small scale in woodlands accessible to urban charcoal traders. Currently very little investment (in the form of hand tools, such as axes, hoes, shovels and forks), is required to enter the charcoal production industry.

Tree felling and kiln building. The wood used in charcoal production is either clear or selectively cut with hand axes. Selection is based on species and/or tree size. A high proportion (over 90 per cent) of uncut stems are small (<10 cm dbh – diameter at breast height). The cord wood (short lengths suitable for firewood rather than building) in uncut stems represents about 10 per cent of the pre-felling stock (Chidumayo 1991b) and most of this consists of species whose wood is either too hard to cut or produces charcoal that supposedly sparks or has a high ash content.

Trees are stumped at about 30 cm above ground, and boles (trunks) and limbs cross-cut into 1–2 m logs either immediately or after air-drying for up to several months (Ranta and Makunka 1986); fresh logs may also be air-dried before kiln building. Logs are then hauled to a selected kiln site within the cut-over area. The moisture content of the kiln wood ranges from 24–62 per cent (Hibajene 1994). The kiln is made by piling logs cross-wise on stringers up to a height of 1.5–3.0 m. Occasionally logs are piled lengthwise. The length of rectangular kilns ranges from a few metres to 85 m with an average stacked volume of 51 m^3 and a range of 5–88 m^3 (Chidumayo and Chidumayo 1984). The top of the kiln is covered with a layer of grass or leafy twigs before covering the whole kiln with soil lumps meticulously dug around its perimeter. The thickness of the soil wall covering the kiln is about 40–45 cm on the sides and 20–25 cm on the top (Hibajene 1994).

Wood carbonization and kiln tending. The kiln is ignited through a hole on one of the narrow sides and the hole sealed off with soil once carbonization has commenced. During carbonization the wood in the kiln passes through four sequential stages: dehydration, combustion, charring and cooling. The time required for each phase depends on the size and moisture content of the wood, the skill of the producer and weather conditions. The kiln temperature may rise to 500–700 °C during carbonization (Boutette and Karch 1984; Ranta and Makunka 1986). When the wood is charring, the volume of the kiln drops to half its original size (Ranta and Makunka 1986), and this causes the soil wall to collapse with consequential smoke emissions. The degree of kiln tending during the carbonization process determines charcoal yield rate.

Kiln tending involves two main activities. Firstly, when carbonization rate is too slow the producer makes holes in the soil wall to allow more air into the kiln. This accelerates the carbonization process, but the holes must be sealed off once the desired rate of carbonization is attained to avoid losses through ashing. Secondly, when the wall of the kiln collapses, timely repairs must be made to avoid charred logs burning to ashes. Experienced charcoal producers can intuitively assess the carbonization rate and will take the necessary tending measures to optimize charcoal yield.

Kiln breaking and charcoal recovery. Once carbonization is completed, the temperature of the kiln drops. This cooling permits the producer to break the kiln and extract the charcoal. In practice, large kilns are broken intermittently and small amounts of charcoal removed with a large garden fork before the charcoal has cooled sufficiently to be packaged in bags. To avoid the hot charcoal catching fire, it is covered with soil to complete the cooling process. On average a regular bag of charcoal contains 41 kg ($SD = 0.5$) of charcoal (Chidumayo and Chidumayo 1984; World Bank 1990). After kiln breaking and recovering the charcoal, the kiln site is referred to as a charcoal spot (Lees 1962) because it consists of a mixture of burnt soil and un-bagged charcoal (i.e., charcoal soil), largely in the form of charcoal fines. Charcoal spots cover 2–3 per cent of the cut-over area (Chidumayo 1993b).

Because charcoal production involves the use of about 94 per cent of the AG cord wood biomass (Table 2.11), it is a significant cause of deforestation in miombo woodland. In 1990 at least 41 700 ha were cleared for urban woodfuel in Zambia. Eighty per cent of this deforestation was in the Central,

Table 2.11 *Cord wood biomass utilization for charcoal production in central dry miombo in Zambia*

Biomass class	Quantity (oven-dry t ha-1)
Total cord wood	75.4
Uncut cord wood	4.3
Cut cord wood	71.1
Charcoal recovered	16.6
Uncarbonized wood	2.0

Based on Chidumayo (1991b)

Table 2.12 *Temporal changes in the rate of deforestation in the Zambian Copperbelt*

Period	Clear-felled area (ha)	Mean annual deforestation (ha)
1937–46	9 513	951.3
1947–51	56 250	11 250.0
1952–56	70 650	14 130.0
1957–61	14 000	2 800.0
1962–84	89 436	3 888.5
1937–84	239 849	4 996.9

After Chidumayo (1989b)

Wild foods

Copperbelt and Lusaka Provinces as can be determined from Table 2.10. During 1937–84 nearly 240 000 ha of miombo in the Zambian Copperbelt were deforested by woodfuel harvesting (Chidumayo 1989b; Table 2.12).

Miombo is a source of many plant food resources which include leaves, fruits and roots. Many of the edible leaves are of herbaceous plants, some of which grow as weeds in cultivated areas and fallows. The most common herbaceous vegetables include *Amaranthus hybridus, Cleome gynandra, Corchorus olitorius, Sesamum angustifolium, S. angolense* and *Bidens pilosa*. Others are *Amaranthus spinosus, A. thunbergii, Celosia trigyna, Portulaca oleracea, Cleome monophylla* and *C. hirta* (Vernon 1983). The few edible tree leaves in miombo include the tender foliage of *Afzelia quanzensis* and *Fagara chalybea*. Miombo produces a variety of edible fruits which have been exploited by man in Zambia since the Upper Pleistocene period (Musonda 1986). Eighty-two per cent of the 33 Zambian edible wild fruits listed by Chidumayo and Siwela (1988) are found in miombo, with the greatest diversity in wet miombo (Table 2.13).

Table 2.13 *Diversity and abundance of wild fruit trees in Zambian miombo*

Diversity/Abundance	Dry miombo	Wet miombo
Mean species/0.04 ha:		
fruit species	3.8	6.4
total species	18.0	22.4
Mean stems/ha:		
fruit species	96.7	186.0
total species	626.7	764.1
Mean stems/fruit species	10.2	17.9

Based on Chidumayo and Siwela (1988)

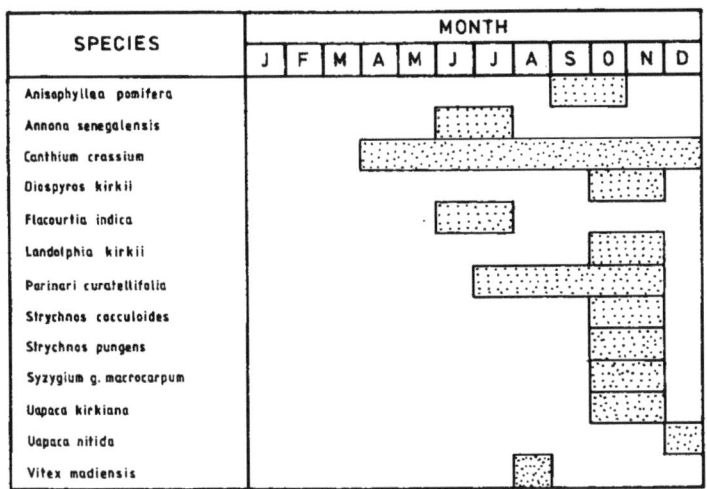

Figure 2.8 *Seasonal availability of ripe edible wild fruits in Zambian miombo (Based on Holden 1988 and Musonda 1986)*

The most common edible wild fruits in Zambian miombo are *Anisophyllea* spp., *Parinari curatellifolia* and *Uapaca kirkiana*. The highest diversity of edible wild fruits in Zimbabwe are also found in miombo (Campbell *et al.* 1991). Factors that influence the local utilization of edible wild fruits include species distribution, abundance, flavour and tradition.

For the majority of species, the ripe fruits are available during the hot dry season (Figure 2.8), but little is known about the quantities of fruit produced, yearly variability and losses due to damage. *Uapaca kirkiana* shows yearly variations in fruiting. Chidumayo (1993a) found that 14 per cent of trees fruited in 1990 compared to 19 per cent in 1991 and 1992, while only 5 per cent fruited in 1993 at two central dry miombo sites in Zambia. The Zambia National Council for Scientific Research found that the fruiting of *U. kirkiana* in other parts of the country was low in 1986 compared to 1985 (unpublished), but fruit production varied with soil type from 0.24–1.1 t ha^{-1}.

During the rainy season, miombo produces up to 25 species of edible mushrooms (Pegler and Piearce 1980). The most common and popular species are the termite mushroom, *Termitomyces letestui*, the christmas mushroom, *Amanita zambiana* and chanterelles, *Cantharellus* sp. These species show differential fruiting phenology and are therefore available at different times of the rainy season. Chanterelles consistently appear from December to February, but the availability of termite and christmas mushrooms may vary from year to year. A three-year study revealed that the termite mushroom was available during November 1990, November 1991–January 1992 and November 1992–December 1993 in dry miombo in central Zambia (Chidumayo 1993a). During the same period, the christmas

Table 2.14 Reputed miombo woodland medicinal trees

Species	Part used	Use
Afzelia quanzensis	Bark	Relieves toothache
Albizia antunesiana	Root	Prophylactic against colds and coughs
Cassia abbreviata	Bark	Antibiotic
	Root extract	Relieves toothache
Combretum molle	Leaf paste	Treatment of wounds and sores
Dichrostachys cinerea	Bark powder	Treatment of skin ailments
	Fresh leaves	Treatment of wounds and sores
Diospyros mespiliformis	Crushed root	Ringworm treatment
	Crushed shoot	Treatment of wounds and sores
Diplorhynchus condylocarpon	Pounded bark	Wound dressing
	Chewed leaves on forehead	Relieves headache
	Root extract	Cough remedy
Garcinia huillensis	Bark infusion	Aphrodisiac
Hymenocardia acida	Vapour from boiling leaves	Relieves headache
Kigelia africana	Ripe fruit	Purgative
Lannea stuhlmannii	Leaf paste	Wound and sore dressing
Piliostigma thonningi	Chewed fresh leaves	Cough relief
Pterocarpus angolensis	Bark paste/ash	Treatment of skin ailments
Rothmannia whitfieldi	Unripe fruit juice	Treatment of wounds and rashes
Strychnos innocua	Seed	Emetic properties

Based on Storrs (1982)

mushroom was available in December 1990, December 1991–March 1992 and December 1992–January 1993, with the highest production in the 1992–93 season.

Edible roots in miombo are largely confined to herbaceous plants. The most common and popular of these in Zambia are the tubers of the orchid, *Satyria siva*, wild yam, *Discorea hirtiflora*, and the legume *Rhynchosia insignis*. The orchid occurs in dambos within the miombo ecosystem, the legume is a woodland plant and the wild yam occurs on termite mounds. The orchid tubers are used to prepare a thick jelly, locally called *chikanda*, which is eaten as a relish, while the roots of *R. insignis* are used to prepare a sweet beverage, locally called *monkoyo*. These products are also widely sold throughout the country.

Little is known about the impact of harvesting food plant resources in miombo. The harvesting of mushrooms probably has the least impact. Fruits for subsistence consumption are usually collected on the ground after the ripe fruits have fallen. However, with the emergence of wild fruit markets, tree cutting to collect fruit has increased (Chidumayo 1993a).

This practice affects woodland structure and reduces the stock of fruiting trees.

The harvesting of roots is equally damaging to plants. Root biomass in herbaceous plants accounts for 55–70 per cent of total plant biomass (Chidumayo 1993a). Traditionally only lateral roots of *R. insignis* were harvested and the remaining tap root provided the resources for the survival of the plant and the regeneration of lateral roots. With the emergence and expansion of *R. insignis* root markets, commercial exploiters harvest both lateral and tap roots. Such over-exploitation can reduce the survival rate of these plants.

Medicinal plants

Many miombo trees and herbs are used for a variety of medicinal purposes. Among the medicinal trees, the most common parts used are leaves, bark and roots. Of the 106 trees whose medicinal parts were documented by Storrs (1979), the frequency with which the various parts are used ranges from 4 per cent for wood, 9 per cent for fruit and seed, 50 per cent for leaves, 66 per cent for bark and 74 per cent for roots. Table 2.14 is a list of the most reputable medicinal trees in miombo and their uses. However, little is known about the frequency and extent of their use, and the significance of medicinal plants in Zambia.

Leaf biomass in miombo constitutes about 3 per cent of total woody plant biomass (Table 2.15), and rarely does leaf harvesting for medicinal purposes result in plant mortality. In contrast, bark and root harvesting can damage plants and lower their survival rate. In fact ring-barking is often used as a technique for killing trees without cutting them. It is not uncommon in areas of high demand to find medicinal trees which have died due to debarking and root harvesting. In miombo, the fatal debarking of *Pterocarpus angolensis* and *Cassia abbreviata* (Table 2.14), especially in and around urban areas, has been reported (Chidumayo 1993a).

Table 2.15 *Structure of woody plant biomass in old-growth dry miombo in central Zambia*

Biomass class	Biomass	
$(t\ ha^{-1})$	(% of total)	
Above ground:		
wood	47.8	45.1
bark	16.5	15.6
leaves	2.9	2.7
Below ground:	38.8	36.6
Total	106.0	100.0

Based on Chidumayo (1993b and Edmonds (1964)

Animal products

Secondary products from miombo include direct animal products, such as meat and edible insects, and those deriving from the activities of secondary producers, such as honey.

Bush meat

Zambia has 19 national parks and 34 game management areas (GMAs) which cover 8.4 per cent and 22 per cent of the country respectively (Figure 2.9). Miombo is the most extensive vegetation type in national parks and GMAs. Ecosystems are legally protected from any form of destructive use in national parks, while killing game for consumption is regulated through licensing in GMAs. Legal hunting includes licensed safari (by foreigners), resident (by local hunters in GMAs) and non-resident (by hunters resident in Zambia but living outside GMAs) and game control hunting, while illegal hunting includes all forms of unlicensed hunting, including subsistence and commercial poaching.

Most wildlife areas in Zambia have a long history of traditional hunting for meat (Marks 1976), but the decline in wildlife species only became evident following an upsurge in commercial poaching during the late 1970s. Nevertheless, small game, such as duiker and small mammals (rodents and hares), and birds, such as francolins and guineafowl, occur widely in miombo outside national parks and GMAs, and are a common source of bush meat for subsistence needs in many parts of the miombo region.

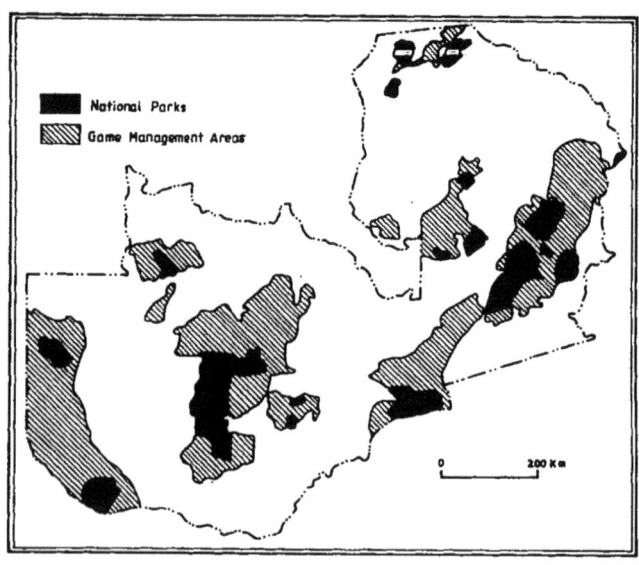

Figure 2.9   *Distribution of national parks and game management areas in Zambia*

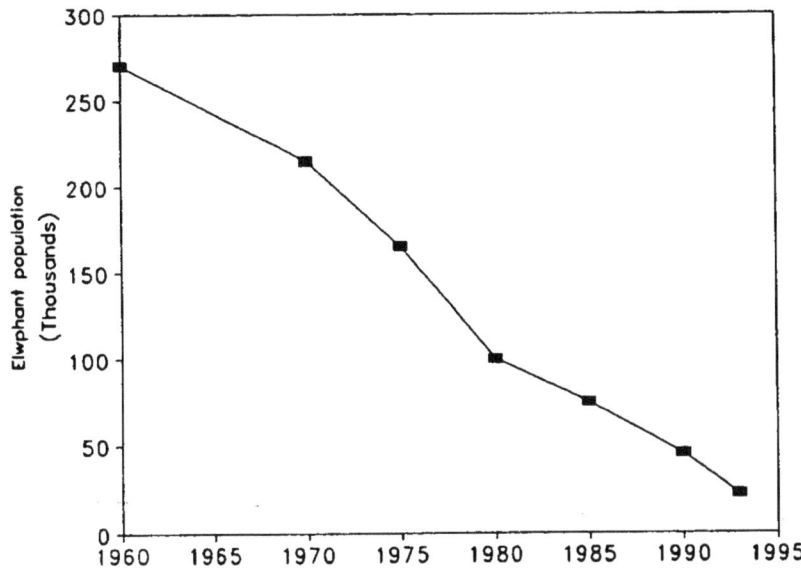

Figure 2.10   *Decline in the elephant population in Zambia during 1960–93*

Although the populations of large mammals, such as elephant, have been greatly reduced as a result of over-hunting (Figure 2.10), these large mammals can cause considerable damage to vegetation. In the Malawian dry miombo in Kasungu national park, elephant browsing reduced the rate of woodland growth while stimulating the proliferation of stems in the 2–4 m height class (Bell and Jachmann 1984). Elephant damage was also responsible for changes in the numbers of *Brachystegia boehmii* in Sengwa wildlife area of Zimbabwe during 1972–6 (Guy 1981). The density of live trees in Sengwa miombo decreased from 321 ha^{-1} in 1972 to 247 ha^{-1} in 1976, while shrub density increased from 3102 in 1972 to 5093 ha^{-1} in 1976.

Edible insects

The most common edible insects in miombo are caterpillars and termite alates. Termite alates, especially of *Macrotermes* spp., are dispersed during the rainy season and are captured and eaten as a source of protein. Termites in miombo are both herbivores and decomposers (Malaisse *et al.* 1975) and therefore play a significant role in nutrient cycling also (see Figures 1.10 and 3.4).

Miombo produces many edible caterpillars which are valued in many parts of the miombo region (Holden 1988; Lees 1962; Malaisse 1978). The caterpillars of the edible emperor moth, *Elephrodes lactea*, feed on *Brachystegia, Julbernardia* and *Isoberlinia* leaves and the species occurs widely in miombo. The eggs which are laid on trees in September hatch in October and the young caterpillars feed on the new foliage. The caterpillars grow very fast and moult several times before maturing. They are

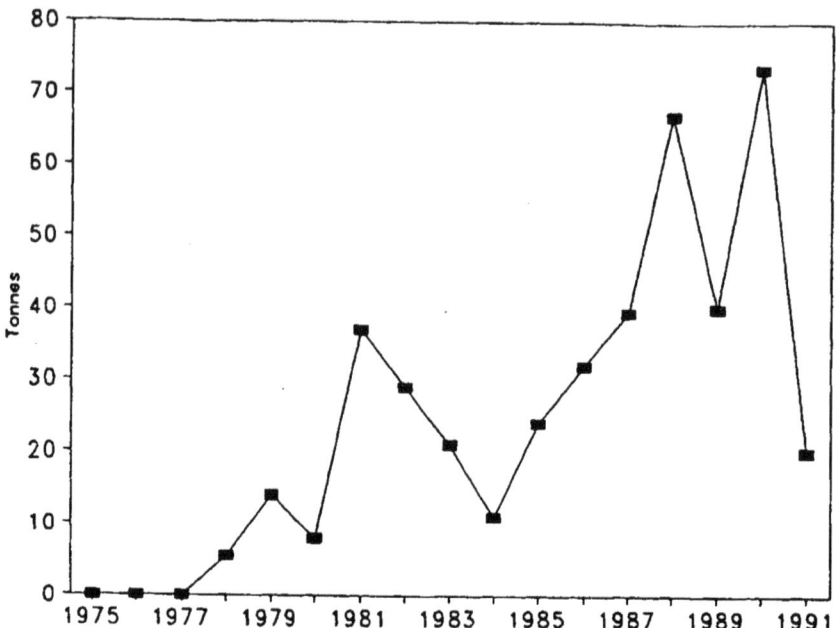

Figure 2.11 *Changes in marketed honey in North-western Province, Zambia, during 1975–91*

harvested during November and December. On a dry-weight basis, the caterpillars contain over 60 per cent protein. In addition to subsistence consumption, local trade in edible caterpillars is growing continuously, especially in urban areas.

Caterpillars are probably among the most important herbivores in miombo, and some are capable of spectacular destruction and total defoliation over large areas. It has been estimated that, at peak density, caterpillars can consume nearly 100 kg of foliage ha^{-1}, but 90 per cent of this is returned to the soil through faeces (Malaisse 1978). This represents a considerable nutrient transfer from tree leaves to the soil. Defoliated trees often produce new replacement leaves within a few months (Campbell 1988).

The main impact on miombo woodland is caused by caterpillar harvesting. Often tall trees are cut to collect caterpillars, and where host trees are gregarious, clearings of up to 2 ha may be made (Lees 1962). This has the same effect on woodland structure as selective wood harvesting. However, because of the high tree density in regrowth miombo, the highest density of caterpillars usually occurs in regrowth miombo in which tree felling is not necessary to harvest caterpillars (Holden 1988).

Honey

Miombo is important for honey production by bees, especially the species *Apis mellifera*. In Zambia the most important honey-producing region is in the North-western Province (Figure 2.11).

Honey collection often involves the cutting of host trees, and traditional beekeeping based on bark hives also damages trees. Bark-hive making involves total ring barking of a portion of the bole, which can cause mortality. In North-western Province up to 435 000 trees per year are destroyed for bark-hive making (Clauss 1992). Although this destruction is confined to large trees of certain species, saplings are also destroyed by tearing off bark strips for cordage used in hive making. The impact of this activity on woodland is similar to that of selective wood harvesting discussed earlier in the chapter.

CHAPTER 3

The Role of Fire

FIRE IS ONE of the most important ecological factors affecting miombo. The use of fire in the miombo region dates back to the Early Stone Age, about 60 000 years ago (Phillipson 1971; Clark 1975). Evidence obtained at the old Kalambo prehistoric site near Mbala in northern Zambia shows that people made deliberate use of fire in Zambia about 53 000 years ago during the Upper Pleistocene. But Stone Age man had used fire for 200 000 years before that for space heating, cooking, honey gathering and hunting.

The occurrence of fire

Climate and flammable biomass determine the occurrence of wild fires. The climate in the miombo region is characterized by a long, dry season (April–November). The highest temperatures and lowest relative humidity occur during the hot, dry season from August to November (Figure 3.1). In central Zambia, average wind speed is highest (13–16 km h^{-1}) during April–October compared to 8–11 km h^{-1} during other months. It is during this period of dry, hot and windy conditions that the risk of fire is greatest.

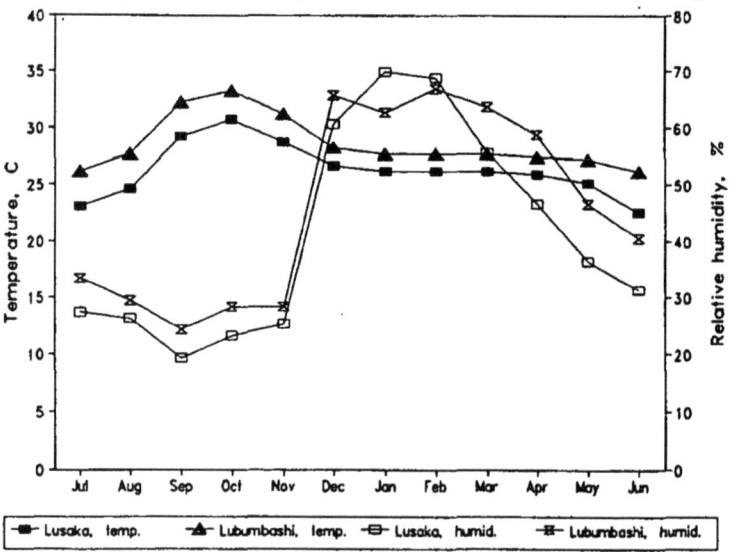

Figure 3.1   *Mean daily maximum temperature and relative humidity in miombo*

The flammable biomass in miombo consists of litter (small twigs and leaves) and standing dead grass and other herbs. In old-growth miombo, flammable biomass ranges from 5–8 t ha^{-1}, of which 70–90 per cent is leaf litter and standing grass (Table 3.1). In young regrowth (under 6 years old) the biomass fuel is 7–12 t ha^{-1}, of which 55–70 per cent is leaf litter and grass and a significant proportion is made up of wood debris, persisting from the time of felling. The biomass fuel structure in older regrowth (6–16 years old) is similar to that in old-growth miombo (Table 3.1). The biomass fuel during the dry season is made up of the following:

(1) undecomposed dead biomass at the end of the previous rainy season;
(2) above-ground grass biomass produced during the previous rainy season; and
(3) current dry-season litter fall.

Annual AG grass production can be estimated by the mean peak standing biomass at the end of the growing season (see Figure 4.9). In dry miombo this averaged at 1150 kg ha^{-1} at four sites over three seasons (Chidumayo 1993a). Mean peak AG grass biomass in wet miombo under a cattle-grazing regime averaged 488 kg ha^{-1} in old-growth and 2006 kg ha^{-1} in cut-over paddocks over four seasons (Hood 1972). Grazing removed about 60 per cent and 50 per cent of the peak biomass in old-growth and coppice plots by the end of the dry season.

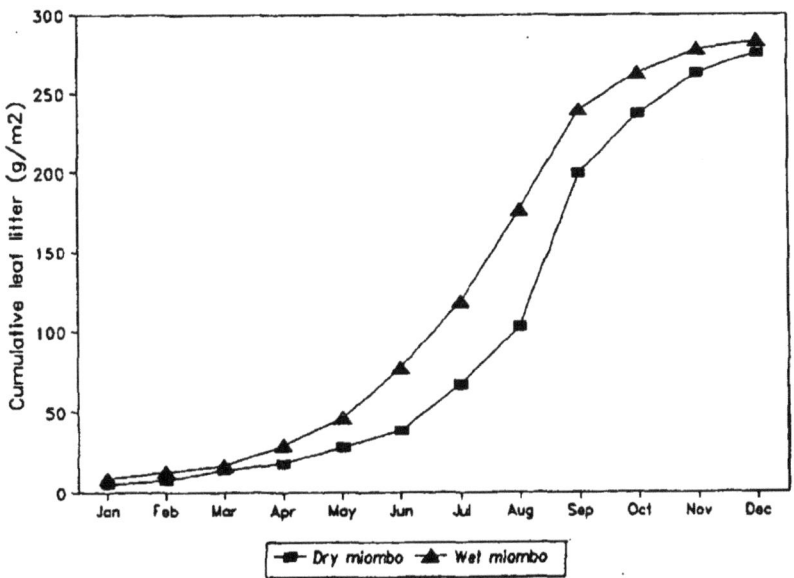

Figure 3.2 *Cumulative tree leaf litter fall in miombo (Based on Chidumayo 1993a and Malaisse et al. 1975)*

Table 3.1 *Above ground flammable biomass (oven-dry kg/ha) in miombo woodland during the dry season*

Biomass class	Dry miombo (1988)			Old-growth (1992)	Wet miombo	
	2-yr coppice	6–16-yr old regrowth	Old-growth		4-yr coppice (1992)	Old-growth (1992)
Litter	1 880	2 250	3 120	3 092	1 490	3 806
Standing grass	6 620	1 610	2 640	671	2 462	1 187
Pod valves	0	90	210	–	–	–
Wood	3 800	1 340	2 150	1 249	3 165	598
Total	12 300	5 290	8 120	5 012	7 117	5 591

Based on Chidumayo (unpublished) for 1988 data; Shea et al. (1993) for 1992 data

The temporal dynamics of leaf litter in miombo is shown in Figure 3.2. Maximum leaf litter fall occurs in August and September, and cumulative litter biomass reaches about 2.5–3.0 t ha^{-1} at the end of each year. This implies that the amount of leaf litter biomass burnt depends on the time of burning. More fuel is available at the end than the beginning of the dry season.

The annual wood-litter fall in miombo is small. In dry miombo, wood-litter fall is about 31 kg ha^{-1} in old-growth, and 8 kg ha^{-1} in young regrowth (Chidumayo unpublished). However, in old-growth wet miombo, annual wood-litter fall over five years averaged 830 kg ha^{-1} in southern Zaire (Malaisse *et al.* 1975). The fall of pod-valve litter varies from year to year (Malaisse *et al.* 1975; Campbell *et al.* 1988). Chidumayo (unpublished) found an average pod-valve litter fall of 88 kg ha^{-1} in old-growth dry miombo, and none in young regrowth. However, because both wood and pod-valve litter decompose very slowly, it remains on the woodland floor for several years (Campbell *et al.* 1988). Over one rainy season at four dry miombo sites, the mass lost from a layer of pod-valve litter averaged 23 per cent, compared to 77 per cent for tree leaf litter (Chidumayo unpublished).

The impact of fire on biomass fuel depends on the time of burning, which is related both to the moisture content (MC) and the amount of fuel. The amount of biomass fuel in miombo increases as the dry season progresses (Figure 3.2). The MC of the dry standing grass and tree leaf litter is lowest (under 7 per cent of dry weight) from August to October (Figure 3.3). In the miombo region, this is also the period when the highest mean daily temperatures and wind speed and lowest relative humidity are experienced. These are favourable conditions for high-intensity fires with a high combustion factor and rapid spread. Average flame height is 2.5 m and the rate of fire spread is 10–60 cm per second (Table 3.2). The combustion factor (percentage of biomass fuel consumed) of miombo woodland fires is typically 100 per cent for grass and 80–90 per cent for leaf litter, but is extremely variable (0–70 per cent) for wood litter (Shea *et al.* 1993).

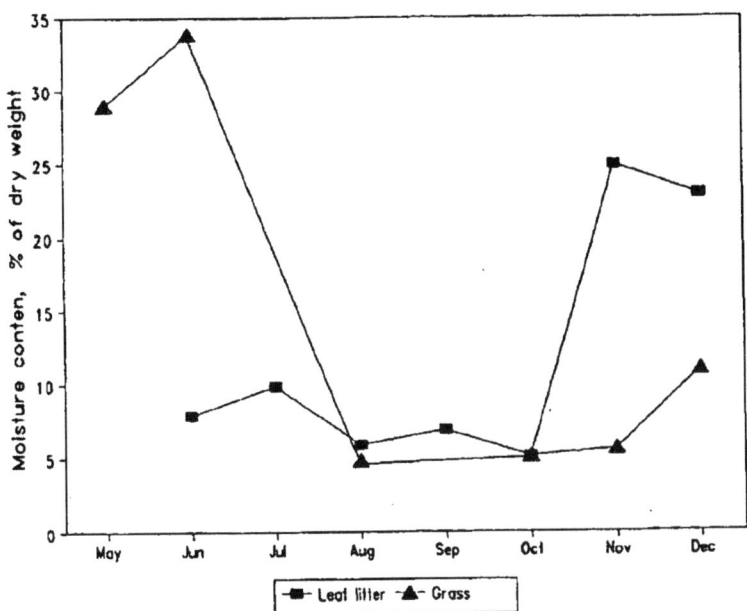

Figure 3.3 *Seasonality in mean moisture content of standing dry grass and tree leaf litter in dry miombo*

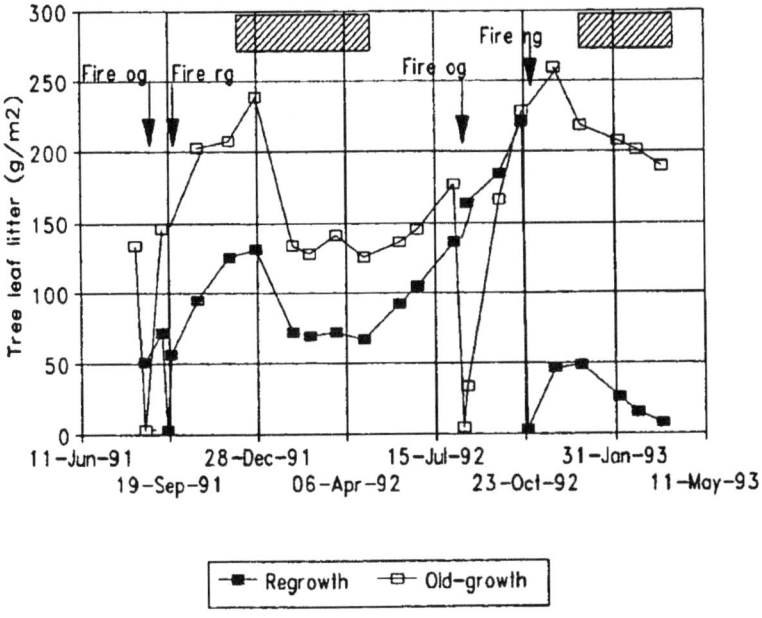

Figure 3.4 *Effects of decomposition and fire on the accumulation of tree leaf litter in dry miombo. Shaded bars indicate periods of decomposition.*

Table 3.2 *Some characteristics of wild fires in miombo woodland during August and September*

Parameter	Mean values
Flame height (m)	2.4
Rate of spread (m/second)	0.3
Biomass fuel consumed (%):	
grass	100.0
tree leaf litter	89.0
wood litter	38.0

If a fire occurs before tree leaf fall is completed, leaf fall is accelerated immediately after the fire, and the pre-fire leaf litter biomass may be restored (Figure 3.4). This is useful in protecting the soil surface from solar radiation and raindrops at the onset of the rainy season. Since grass is the common fuel used to ignite fires in miombo (Shea *et al.* 1993), its complete combustion during a fire reduces the probability of the post-fire leaf fall being burned again in the same dry season. However, if the fire occurs after leaf flush, there is no input of leaf litter because the scorched new foliage remains attached to twigs in the canopy. This has two main ecological consequences. First, the bare soil is subjected to direct solar radiation and the impact of raindrops at the beginning of the rainy season, prior to the development of an adequate herbaceous cover. Second, there is inadequate surface litter for decomposition and the transfer of nutrients to the top-soil during the rainy season.

The causes and frequency of fires

The evolution of agriculture and iron manufacturing during the Iron Age marked the beginning of the use of fire for land clearing prior to cultivation and livestock grazing and for making charcoal for iron smelting. Traditional uses of fire today have changed little from those of our ancestors. Wild fires are also used to clear bush and undergrowth to improve visibility around settlements, foot paths and roads, and to keep away dangerous animals, such as snakes and carnivores (Table 3.3).

Table 3.3 *Perceptions about the causes of wild fires among 1410 rural households in the miombo region of Zambia*

Cause/Use of fire	Positive responses
Charcoal production	1
Improving landscape visibility	5
Keeping away dangerous animals	28
Management of grazing	41
Preparation of land for cultivation	74
Honey collection	76

After Chidumayo (1986)

In the Zambian Copperbelt the monitoring of 490 plantation fires during 1976–84 revealed that 14 per cent were caused by hunters, 8 per cent by charcoal producers, 5 per cent by plantation workers, 3 per cent by lightning and 6 per cent were caused by stray fires spreading from adjacent miombo woodlands (Chidumayo 1989b). Although the cause of 64 per cent of the fires could not be determined, natural fires appear to be rare in miombo woodland. The majority of wild fires are therefore man-made. The majority of miombo woodland fires occur during the hot, dry season. Chidumayo (1993a) monitored the occurrence of fire at four dry miombo sites in central Zambia over a four-year period (1990–3). Out of the 13 fires at the sites, 15 per cent occurred in August, 39 per cent in September and 46 per cent in October, and the mean fire return period was 1.6 years. Long-term data on forest plantation fires in the Zambian Copperbelt support the observation that the hot, dry season (August–October) has the highest fire frequency in the miombo region (Figure 3.5).

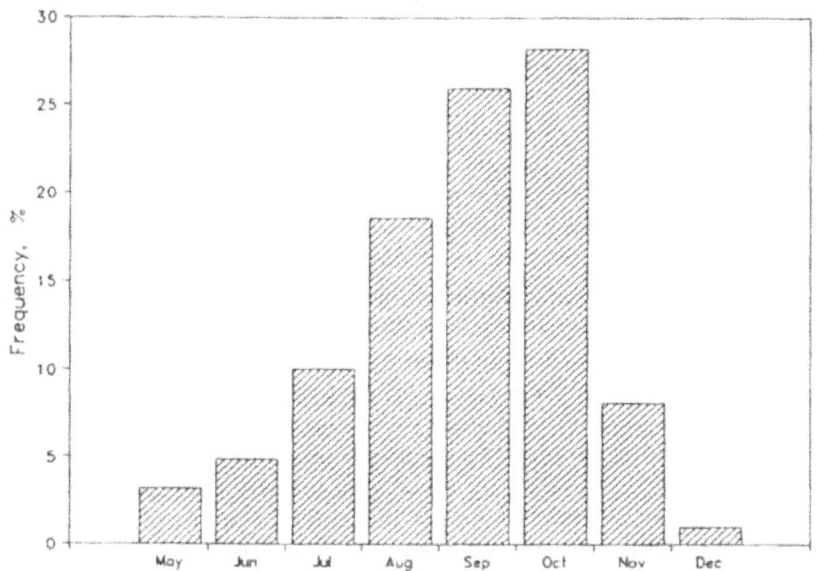

Figure 3.5 *Mean monthly frequency of 12003 forest plantation fires in the Zambian Copperbelt during 1975–93*

The pattern of wild fires in miombo is closely linked to both the fire environment and cultural and socio-economic activities of the miombo inhabitants. Traditionally bush burning is discouraged until crops have been harvested because of the danger fires pose to crops, while the first rains usually mark the end of the burning period. In 50 per cent of the years, the first rains will have occurred by the last week of October in Zambia. Consequently, traditional bush burning rarely extends into November (Figure 3.5).

In the chitemene shifting cultivation region of Zambia, the annual cycle of chitemene activities impose additional restrictions on the bush burning period. The burning of wood piles for chitemene occurs close to the onset of the rainy season, in September and October. Bush burning is discouraged during the early dry season, initially to protect crops and later in the dry season to protect chitemene wood piles until the kindling season in September–October. Thus the hot, dry season (August–October) coincides with the traditional bush burning period in Zambia.

The extent of each wild fire in miombo has not been adequately studied. Chidumayo (unpublished) mapped the extent of 13 fires at four dry miombo sites over the 1990–3 period. On average, each fire burnt 75 per cent of the site, regardless of the month the fire occurred during August–October.

Plant adaptations to fire

The impact of fire on miombo woodland depends on the time and frequency of burning and on the amount of flammable biomass. A number of attributes enable miombo woodland plants to survive fire.

When seeds are buried, they are less susceptible to damage by fire than mature plants. Annual plants in miombo woodland usually survive fire through seeds buried in the soil. But the majority of miombo woodland trees do not have soil seed banks and seeds dispersed on the soil surface and in the litter are more susceptible to damage by fire than buried seeds. However, seeds protected by indehiscent fruits are protected from the direct effects of fire. Such species include the winged fruits of members of *Combretum, Terminalia* and *Pterocarpus,* which disperse their fruits in the dry season. The woody fruit of *Gardenia subacaulis*, a geophytic suffrutex, protects the seeds from fire and only releases them during the rainy season after the fruit decomposes (Medwecka-Kornas 1980).

Actively growing or reproducing plants are more susceptible to fire damage than dormant ones. By scheduling growth and reproductive activities when the risk of fire is lowest, plants minimize the damage caused by fire (Frost 1985). Many herbaceous and some woody plants in miombo are dormant during the dry season when the risk of fire is greatest.

The AG parts of the majority of herbaceous plants in miombo die during the dry season and perennating buds are either BG (true geophytes) or at ground level (hemicryptophytes). These renovating buds are protected by old leaf or stipe bases and are therefore not killed by annual fires. This fire protection method is found in perennial bunch grasses (*Graminae*), sedges (*Cyperaceae*) and ferns (*Pteridophytes*), and serves as an example of convergent evolution among these families of plants.

The majority of ferns and sedges in Zambian miombo are fire resistant with a long dry-season dormancy period and a short active phase (Kornas 1978; Medwecka-Kornas and Kornas 1985). Often the fire leaves charred old culms and leaves or leaf bases protruding above the ground level which protect perennating buds. Dormancy is broken by fire which stimulates shoot development in a few weeks following a fire, regardless of the time of burning (Medwecka-Kornas and Kornas 1985).

Most woody suffrutices in miombo have poor shoot growth in the absence of fires. In such plants, which include *Annona stenophylla*, *Gardenia subacaulis* and *Lannea edulis*, fire stimulates the production of vigorous new shoots and flowers after killing the slightly lignified or herbaceous AG shoots (Medwecka-Kornas 1980; White 1976). These plants have strong woody BG rhizomes which serve as perennating organs just below the soil. Although White (1976) has argued that the geoxylic suffrutex growth-form, with massive woody BG axes and annual or short-lived AG shoots, evolved in response to extreme nutrient deficiency (oligotrophy) and waterlogging conditions, nonetheless these features offer effective protection against fire, and may therefore represent pre-adaptations to fire (Medwecka-Kornas and Kornas 1985).

Fire tolerance among miombo trees consists of three attributes: the

Table 3.4 *Leaf flush and dry season wood moisture content (% dry weight) in common dry miombo woodland trees in central Zambia*

Species	Percent trees with leaf flush				Wood mean moisture content (%)
	Aug 1991	Sep 1991	Oct 1991	Nov 1991	
Albizia antunesiana	0	13	23	86	77
Brachystegia boehmii	0	1	100	100	81
Brachystegia spiciformis	33	50	100	100	70
Brachystegia utilis	0	0	100	100	61
Bridelia carthatica	0	0	100	100	69
Burkea africana	0	0	100	100	82
Dichrostachys cinerea	0	0	91	100	39
Diplorhynchus condylocarpon	0	0	100	100	84
Faurea saligna/F.speciosa	0	0	60	72	76
Isoberlinia angolensis	19	45	96	100	82
Julbernardia globiflora	2	41	97	100	65
Monotes africanus	0	0	97	100	68
Ochna schweinfurthiana	0	0	100	100	85
Pericopsis angolensis	0	0	100	100	80
Phyllocosmus lemaireanus	0	7	93	100	78
Protea homblei/P.gaguedi	2	10	96	100	107
Pseudolachnostylis maprounefolia	0	0	100	100	78
Swartzia madagascariensis	0	8	100	100	58
Uapaca kirkiana	0	3	98	100	100
Uapaca nitida	0	5	100	100	90
Syzygium guineense macrocarpum	0	25	94	100	120

Based on Chidumayo (unpublished)

scheduling of active phenological phases when the risk of fire is lowest, the ability to resist fire by having vital tissues insulated from high fire temperatures, and the capacity to recover vegetatively when fire has damaged plant tissues.

The seedlings of woody plants are particularly susceptible to fire damage. It has been proposed that cryptogeal germination (with the plumule drawn below ground) improves seedling protection from fire by having meristems below the soil surface (Frost 1985). In miombo woodland, this form of germination occurs in *Combretum* and *Pterocarpus* species. Hypogeal germination (with cotyledons at the soil surface) is also said to enhance seedling survival and is found in *Isoberlinia angolensis, Pterocarpus angolensis, Uapaca kirkiana* and *Bauhinia petersiana*. In spite of these germination strategies, cryptogeal and hypogeal species show no differential survival under a fire regime. Seedlings of *U. kirkiana* rarely survive a fire while *I. angolensis* seedlings are rarely killed by fire (Chidumayo 1993a). Furthermore, the dominant *Brachystegia* and *Julbernardia* species in miombo have epigeal germination (with cotyledons above ground). It is therefore difficult to appreciate the selective advantage of cryptogeal and hypogeal germination as mechanisms for fire protection in miombo.

In a fire risk environment, such as miombo, the scheduling of active phenological phases when the fire risk is lowest improves the survival rate and productivity of trees. Leaf flush is a very important phase in deciduous plants, and its timing affects plant growth and productivity. The leaf flush phase of the majority of dry miombo trees occurs in October and November, but in a few species, such as *B. spiciformis, I. angolensis, J. globiflora* and *Protea* spp, leaf flush starts in August (Table 3.4). These species with an early leaf flush are more susceptible to fire damage throughout the hot dry season than those that flush late. For example, whereas August fires have little effect on *I. angolensis*, September fires reduce the productivity of this species. In contrast, both August and September fires have little effect on the productivity of *U. kirkiana*, which flushes late (Chidumayo 1993a: Figure 3.6). Repeated late dry-season fires can therefore damage trees with early leaf flush while selectively favouring those that flush late.

The ability to resist fire among miombo trees is also achieved by a thick bark and/or a high wood MC. On the basis of wood MC, miombo trees can be divided into three groups: those with high MC (> 90 per cent), those with moderate MC (60–90 per cent) and those with low MC (< 60 per cent). The majority of miombo trees have moderate wood MC (Table 3.4). Since the flammability of biomass depends on its MC, species with a high MC, such as *Protea* spp. *Uapaca* spp. and *Syzygium guineense macrocarpum,* are less likely to be damaged by fire than species with a low MC, such as *Dichrostachys cinerea* and *Swartzia madagascariensis. D. cinerea* is an invasive bush in the absence of fire and fire is used to control such bush

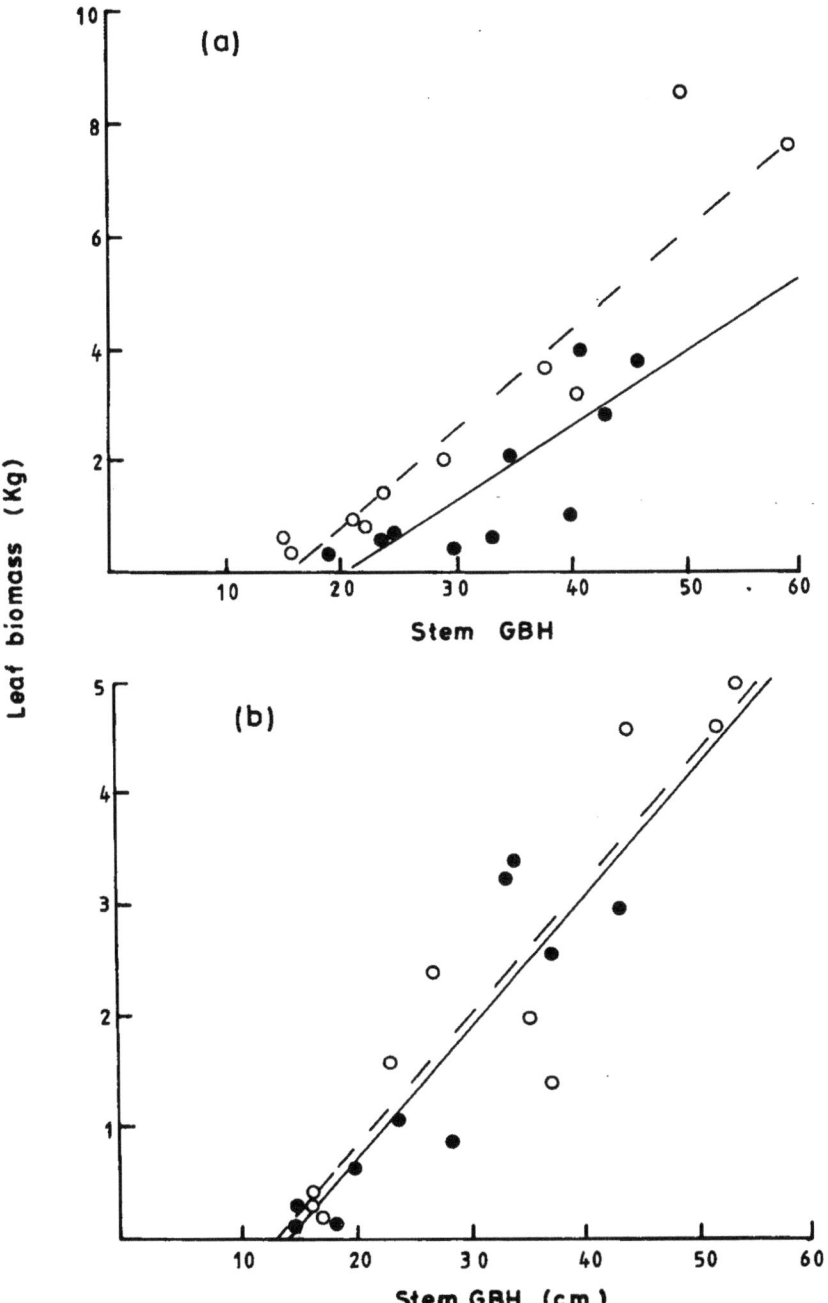

Figure 3.6 *Effect of August (circles) and September fires (dots) on leaf production in Isoberlinia angolensis (a) and Uapaca kirkiana (b)*

Table 3.5 *Effect of fire on trees at a 1988 selectively cut dry miombo site.*

Species	Number of live stems		Mortality (%)
	1988	1991	
Brachystegia boehmii	11	5	18
Dichrostachys cinerea	6	0	100
Isoberlinia angolensis	10	10	0
Julbernardia globiflora	27	11	20
All species	54	26	52

Based on Chidumayo unpublished

Table 3.6 *Average bark thickness of some dry miombo trees in central Zambia*

	Average bark thickness (mm)	
Species	Small stems (< 15 cm gbh)	Large stems (> 15 cm gbh)
Afzelia quanzensis	4.2	–
Albizia antunesiana	3.2	11.4
Brachystegia boehmii	5.2	17.5
Brachystegia spiciformis	4.2	14.0
Burkea africana	2.8	13.5
Dalbergiellea nyassae	3.0	–
Diplorhynchus condylocarpon	4.2	–
Faurea speciosa	5.3	–
Isoberlinia angolensis	4.2	12.6
Julbernardia globiflora	2.6	12.6
Monotes spp.	2.6	10.0
Ochna schweinfurthiana	5.3	–
Parinari curatellifolia	4.8	14.6
Pseudolachnostylis maprounefolia	1.3	8.1
Swartzia madagascariensis	3.6	–
Syzygium guineense macrocarpum	4.9	–
Uapaca spp.	3.0	10.5

Based on Edmonds (1964) and Chidumayo (unpublished)

encroachment. Sweet and Tacheba (1985) reported a 100 per cent mortality of *D. cinerea* by fire at a savanna site in Botswana. At a selectively cut dry miombo site in central Zambia, fire eliminated *D. cinerea* in a few years (Table 3.5).

Bark thickness is often cited as a protection mechanism against fire (Frost 1985; Stott 1988). A thick bark insulates the vital inner tissues, such as the meristems and vascular organs, better than a thin bark. Bark thickness among miombo trees is variable (Table 3.6). Large stems have a bark that is 3–5 times thicker than that of small stems and this may explain why saplings and small poles are more susceptible to fire damage than large stems. During the sapling phase, *Burkea africana, Julbernardia globiflora, Monotes* spp., and *Pseudolachnostylis maproneifolia* have a thin bark and are therefore likely to be more fire sensitive than the other species (Table 3.6).

The capacity to recover vegetatively when fire damages plant tissues is found in the majority of miombo trees. Trees that suffer partial or complete death of AG parts resprout from dormant buds located in the branches, stem, root collar or roots. As long as the apical buds are active, these buds remain dormant, but once apical bud dominance is broken by fire or cutting, the dormant buds are activated. The extent of resprouting depends on the degree of damage to the apical meristems and the condition of the dormant buds. Complete top-kill results in the most vigorous resprouting, while resprouting is poor in very old or large trees or those with partial top-kill.

The effects on succession

Because of the significance of fire in the Zambezian phytoregion, some attempts have been made to relate fire to plant community successions. The successional hypothesis proposed by Lawton (1978) is based on the progressive reduction in fire intensity and the final exclusion of fire through the development of a closed canopy and a corresponding decrease in the productivity of the herbaceous layer. Apparently these changes are initiated by fire-tolerant chipya woodland (see Chapter 1) species which form a light canopy that creates conditions favourable for the establishment of semi-tolerant *Uapaca* species. The latter species further suppress grass production and fire intensity, thereby facilitating the development of *Brachystegia* and *Julbernardia* miombo canopy species. The *Uapaca* group also protects the regeneration phase of the dry evergreen forest (*mateshi*) trees and shrubs. In the absence of fire, chipya woodland in northern Zambia can therefore be replaced by dry evergreen forest, while early burning at the beginning of the dry season permits the replacement of the *Uapaca* group by wet miombo. Lawton's successional hypothesis was developed on the basis of single observations at different sites in northeastern Zambia, but its validity at any one site has not been demonstrated.

The only long-term fire experiments in wet miombo were started in 1933 at Ndola in the Zambian Copperbelt (Chidumayo 1988b; Trapnell 1959; Trapnell *et al.* 1976). The fire experiments included three treatments: annual early burning in May or June, annual late burning in September or October, and complete fire protection in both old-growth and coppiced plots. Based on 11 years of these different fire regimes, Trapnell (1959) classified the species according to fire tolerance based on their abundance and growth performance under the different fire-treatment regimes. Species most abundant in the late burnt plots were considered fire-tolerant, while those most abundant in the fire-protected plots were considered fire-intolerant and those most abundant in the early burnt plots were considered semi-tolerant (Table 3.7).

Trapnell's fire tolerance classification of miombo trees did not involve statistical comparisons of abundances before and after the experiment, which

Table 3.7 Trapnell's fire tolerance classification of trees and shrubs after 11 continuous years of fire treatments in wet miombo at Ndola in the Zambian Copperbelt

Fire-intolerant/semi-tolerant species	Fire-tolerant species
Brachystegia longifolia *	Anisophyllea boehmii
Brachystegia spiciformis *	Dialiopsis africana
Bridelia carthatica	Diplorhynchus condylocarpon
Bridelia duvigneaudii	Erythrophleum africanum *
Byrsocarpus orientalis	Dombeya rotundifolia
Chrysophyllum bangweolense	Hymenocardia acida
Garcinia huillensis	Maprounea africana
Hexalobus monopetalus	Parinari curatellifolia *
Isoberlinia angolensis *	Pterocarpus angolensis *
Julbernardia paniculata *	Strychnos cocculoides
Lannea discolor	Strychnos innocua
Ochna mechowiana	Strychnos spinosa
Ochna schweinfurthiana	Swartzia madagascariensis
Parinari polyandra *	Syzygium owariense
Pseudolachnostylis maprounefolia	Uapaca nitida
Uapaca kirkiana	Vitex madiensis
Uapaca pilosa	
Xylopia odoratissima	

Asterisk denotes canopy species

may have yielded different results (see Box 4.9). For example, using chi-square statistical comparisons of abundances before and after fire treatments, Chidumayo (1989a) found that although *Albizia adianthifolia, Baphia bequaertii, Marquesia macroura* and *Uapaca kirkiana* at a wet miombo site near Mufulira in the Zambian Copperbelt contributed the largest proportions to dead stems due to fire, these proportions were not significantly larger than the species relative abundances before the experiment, except for *A. adianthifolia* and *M. macroura*. These latter species were therefore considered less fire-tolerant than *B. bequaertii* and *U. kirkiana*.

As has been observed in Chapter 3, the majority of miombo trees have a thin bark when small or have a late leaf flush (60 per cent) and a small proportion has a high wood MC. These are the three main characteristics that enable miombo trees to tolerate fire. None of the miombo trees has a combination of thick bark, late leaf flush and a high MC, nor are there trees with a combination of thin bark, early leaf flush and a low MC. The latter is typical of extremely fire-intolerant species. A low wood MC and either a thin bark or early leaf flush is the only combination that characterizes fire-intolerant species in miombo woodland. These species include *Dichrostachys cinerea* and *Swartzia madagascariensis*. All the other species have either a structural or phenological adaptation to fire. It should be noted that *S. madagascariensis*, which has fire-intolerant characteristics, was classified as fire-tolerant by Trapnell (1959).

Table 3.8 *Effects of long-term (1933–82) fire exclusion and burning on species diversity in wet miombo in Zambia*

Parameter	Fire treatment					
	Early burning plot			No burning plot		
	1933	1982	1933–82	1933	1982	1933–82
Species present:						
canopy	8	10	–	10	11	–
understorey	16	17	–	17	7	–
Species loss:						
canopy	–	–	–2	–	–	0
understorey	–	–	–5	–	–	–10
Species gained:						
canopy	–	–	4	–	–	1
understorey	–	–	6	–	–	1
Persistent species:						
canopy	–	–	6	–	–	10
understorey	–	–	11	–	–	6

Based on Chidumayo 1989a

Successional trends in miombo under fire protection suggest that in the presence of dry evergreen forest the understorey is liable to change due to the establishment of fire-intolerant evergreen lianes, shrubs and juvenile trees (Trapnell 1959). At the Ndola miombo plots, evergreen species, such as *Canthium gueinzii, Dalbergia nitidula, Erythroxylon emerginatum, Marquesia macroura, Parinari excelsa, Rothmannia fischeri, Syzygium guineense* and *Tricalysia congesta,* had appeared in fire-protected plots within 11 years. The shading effect of a closed canopy developed in the fire-protection plots has also caused the loss of heliophytic understorey species, which normally survive under an open or semi-closed canopy maintained by annual burning (Table 3.8).

However, the full subclimax status of wet miombo at the Ndola plots can only be demonstrated if the miombo canopy is replaced by dry evergreen forest. This has not occurred yet (Chidumayo 1988b; Trapnell 1959). Furthermore, such a succession might be possible only in miombo which contains dry evergreen components that are capable of initiating the succession. Succession to dry evergreen forest would therefore be extremely improbable in dry miombo which lacks evergreen elements. Because of these observations, Trapnell (1959) concluded that the *Brachystegia-Julbernardia* canopy in miombo exists not because of fires, but in spite of them.

The most significant and perhaps invariable effect of fire protection in miombo is the enhancement of tree density (Table 3.9). However, such high stem densities under fire protection and early burning are temporary. For example, the density of stems with gbh > 20.0 cm in coppiced plots under early burning and fire protection at the Ndola plots was 95 per cent

Table 3.9 *Changes in tree density after 11 years of fire treatments in wet miombo at Ndola in the Zambian Copperbelt (number ha^{-1})*

Fire treatment	Old-growth plots		Coppiced plots	
	1933 (gbh15 >cm)	1944 (<1.1 m tall)	1933 (gbh>20 cm)	1944 (>0.9 m tall)
Late burnt	403	655	458	455
Early burnt	325	2620	550	2105
Fire protected	480	4143	455	2980

Based on Trapnell (1959)

and 8 per cent of the pre-felling density after 49 years (Chidumayo 1988b).

A structural change observed in the late burnt old-growth wet miombo at Ndola was a succession in the herbaceous layer from one dominated by *Loudetia superba* grass to a tall (1.5–2.4 m) *Hyparrhenia-Andropogon-Loudetia* grassland after five years of treatment (Trapnell 1959). Regular late burning in old-growth wet miombo may therefore induce successional changes in the herbaceous community.

It has been observed that long-term burning in miombo woodland increases soil pH and exchangeable P, Ca and Mg (Trapnell *et al.* 1976). Burning therefore has positive effects on soil nutrient status. Apparently protection from fire does not improve soil nutrient status because termites concentrate organic matter from the surrounding area into their mounds. Consequently, termite mounds in miombo have a higher soil carbon, N, Ca, Mg and K content and pH than top soil in the surrounding area (Trapnell *et al.* 1976).

CHAPTER 4
Management Inventories and Assessments

Inventories

Objectives and procedure

A forest inventory is a system of collecting and analysing data for the purpose of assessing the quantity and quality of stocks of timber and non-timber forest products in a given area. Before embarking on an inventory it is important to seek agreement over the objective of the forest inventory with the users of the information to be collected in order to enhance the quality and usefulness of the inventory. Other issues to be considered include:

(1) The expected output.
(2) Time and financial resources available.
(3) Possibilities of standardization and/or combination with other inventory data.
(4) The forest/vegetation classification system to be used.
(5) The inventory sampling design, including prospects for recurrent inventories.
(6) Availability and use of maps, aerial photographs and satellite imagery.
(7) Availability of personnel and the need for training.
(8) Logistical support.
(9) Field measurements and recording forms.
(10) Possible relationships between variables to be measured.
(11) How data will be analysed and compiled. If computers are to be used in the analysis, some data items may require to be coded. For example, species can be assigned numbers.

The objectives of forest inventories vary a great deal (Bonnor 1987; Palmer and Synnott 1992), but these can be summarized as follows:

(1) Land-use planning which aims at allocating land to different uses.
(2) Forest-development planning which aims at assessing the potential for the development of a particular forest product or products.
(3) Feasibility study for a forest products industry.

(4) Forest-management planning which aims at determining stocks, allocation, utilization and regeneration of forest products and their conservation.

In this handbook, the emphasis is on management inventories which include research inventories that generate useful data for forest management. Most inventories carried out in miombo woodland have focused on stocks of timber and poles. One of the earliest inventories in Zambian miombo was made in the Copperbelt area to assess the stocks of mine timber and poles (Lees 1962). District forest inventories carried out throughout the country during the early 1960s also focused on timber and poles, but a few included cord wood for fuel (Edmonds 1964). Nowadays the multiple use of miombo for energy, grazing, wildlife management, food, medicines and ecological stability (see Chapter 2) has gained greater formal recognition, and forest management is increasingly being called upon to ensure the sustainable supply of both timber and non-timber forest products.

It is not feasible to count and measure all trees, shrubs and herbs in a forest. During a forest inventory, censuses and measurements are made in sample plots, the location, size and number of which depends on the objective of the inventory. A management inventory requires that sample data be applied to larger forest management units. The procedure for carrying out such an inventory includes the following sequential steps.

(1) Define the objectives of the inventory.
(2) Define the total area to be covered by the inventory.
(3) Divide the land in the inventory area into vegetation or land-use types and delineate these on base maps or aerial photographs (Figure 4.1). This makes it easier to determine areas under each vegetation or land-use type, and permits easier control and the execution of inventory work.
(4) Select the preliminary location of sample plots on the map or aerial photographs and determine the number of plots required for each forest type (see Box 4.1).
(5) Carry out a preliminary field survey to determine the suitability of the sample plots selected in (4) and the appropriate plot sizes.
(6) Make adjustments, if any, to (4) and carry out the inventory.
(7) Analyse inventory data from the sample plots by either developing equations or applying them to existing equations to estimate tree basal areas, volumes or biomass. Summarize the analysis by species and/or stem sizes. Compare the plot totals which are then combined with other plot totals for each forest type and average these as basal area/volume/biomass per ha and calculate precision estimates (e.g. standard deviation). Sample standard deviation is calculated as follows:

$$SD = \sqrt{\frac{\Sigma x^2 - (\Sigma x)^2/n}{n-1}} \qquad \text{Standard deviation (1)}$$

where x represents individual sample observations and n is total number of samples.

(8) Combine the area data in (3) with averages in (7) to give estimates for each forest type. Summarize the data (Table 4.1) and prepare a report with maps. Forest management plans can then be prepared from the report and any other available information.

There are two types of forest management inventories: single and recurrent. A single inventory provides information on the current growing stock, while a recurrent inventory monitors growth rates and other changes in the forest through repeat measurements. The design of a recurrent inventory must allow comparisons to be made of the results from successive measurements. In miombo woodland, recurrent phenological measurements should be made either monthly or seasonally, while annual or longer-interval growth re-measurements are recommended because miombo trees grow slowly.

A classification system is important in forest inventories because it permits the subdivision of a heterogenous forest into smaller and more homogenous units. This vegetation stratification also facilitates more efficient sampling and the preparation of maps. Forest classification may be based on composition, size, density and site. If tenure has an effect on management, then forest ownership should be included in the classification criteria.

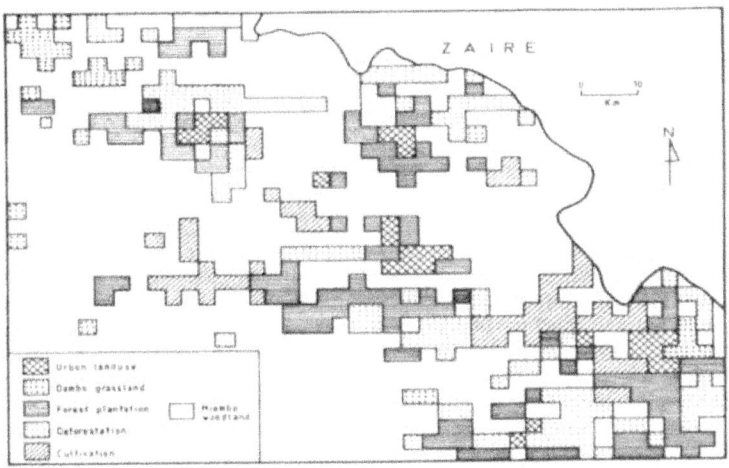

Figure 4.1 *Land use and vegetation covers types in the Zambian Copperbelt in 1984*

Box 4.1 *Comparing differences between random and systematic sampling in a miombo woodland inventory.*

During the workshop to test the draft version of this Handbook, a forest survey was conducted near the Zambia Forestry College in Kitwe, Zambia, in a 164 ha area using random and systematic sampling designs. A preliminary survey of five 20 × 20 m sample plots was made to determine the number of sample plots required to achieve confidence limits (E) of 20% of the mean. The sample data gave a mean (x) basal area (BA) of 21.92 m² ha^{-1} with a variance (S^2) of 58.10. To estimate the number of sampling plots (n) needed for the survey the following formula was used:

$n = (t^2S^2)/(E^2x^2) = (4 \times 58.10)/((0.2)^2(21.92)^2) = 12$, with a variance of 34.8%

A total of 24 sample plots were therefore measured for diameter at breast height (dbh) for all stems with dbh > 10.0 cm, which represented a sampling intensity of 0.6% in the 164-ha forest block. The total forest block was subdivided into 20 × 20 m sampling units using a base map and each unit identified by a grid reference number. Twelve of the sample plots were randomly selected, while the other 12 were systematically selected with an inter-plot distance of 400 m. The location of the sample plots is given in the map shown in Figure 4.2. The analysis of the measurements was as follows:

Sample plot	Mean dbh (cm)		Basal area (m²/ha)	
	Random	Systematic	Random	Systematic
1	18.6	no tree	19.53	0
2	26.6	no tree	19.70	0
3	23.4	25.6	19.13	13.93
4	21.9	28.2	15.53	10.10
5	29.3	27.2	35.28	16.73
6	31.4	19.7	23.93	5.78
7	26.1	29.7	13.48	21.48
8	34.7	28.3	15.78	17.65
9	no tree	39.8	0	30.63
10	18.6	46.4	17.40	28.50
11	27.7	no tree	19.60	0
12	22.6	15.5	10.43	8.78
Mean	25.54	28.93	17.48	12.80
SD	5.111.54	9.38	8.27	10.63
SE		3.13	2.39	3.07

The mean BA for the 24 sample plots was 15.10 m² ha^{-1} with a standard deviation (SD) and standard error (SE) of 9.62 and 1.96, respectively. If this estimate is assumed to be the population mean, then the deviation of the mean for random samples is 2.38 while the deviation of the mean for systematic samples is -2.30. Thus although both sampling designs gave mean BA estimates that were within the range of the assumed population mean of 11.04–19.16 m² ha^{-1} (mean ±tSE), systematic sampling showed more variation among plots because sample plots were distributed both in wooded and open areas (see Figure 4.2). This variability was not adequately captured by random sampling. The use of stratified random sampling would have be more appropriate in capturing this variability. In this exercise, the two sampling designs took almost the same time to execute and therefore it was not possible to determine which of the two designs was cheaper to execute.

However, if the inventory focuses on land uses other than timber production, then relevant criteria should be formulated, defined and used in the classification.

If tenure has an effect on management, then forest ownership should be included in the classification criteria. However, if the inventory focuses on land uses other than timber production, then relevant criteria should be formulated, defined and used in the classification.

Sampling design and plots
The purpose of a sampling design in a forest inventory is to ensure that a desired degree of precision at a minimum or specified cost is achieved. Because of the many factors that determine the selection of a particular design, there is no inventory design that is universally applicable. Among the factors that determine the selection of a sampling design are the following:

Table 4.1 Wood biomass estimates by forest type in Zambia

Vegetation type	Wood biomass (t ha^{-1})			Area (million ha)			Wood stocks (million t)
	Cord	Brush	Total	Total	Dambo grassland	Cropland	
Evergreen forest	158	29	187	3.93	0.49	0.43	562.87
Deciduous forest	58	11	69	1.00	0.13	0.09	53.82
Wet miombo	76	14	90	22.20	2.78	6.66	1 148.40
Dry miombo	58	11	69	13.13	1.64	5.91	385.02
Kalahari miombo	43	8	51	9.76	1.22	0.98	385.56
Munga	46	9	55	3.73	0.47	0.15	171.05
Mopane	46	9	55	4.75	0.59	0.19	218.32
Termitaria	25	5	30	2.77	0.35	0.00	72.6
Grassland	0	0	0	13.01	0.00	1.43	0.00
Aquatic	0	0	0	0.98	0.00	0.00	0.00
Total	–	–	–	75.26	7.67	15.84	2 997.67

Based on Chidumayo (1994a)

(1) The type of information required and its desired precision.
(2) Composition of the forest and its variability.
(3) Landscape characteristics, especially topography and accessibility to and within the forest.
(4) The availability of aerial photographs and/or maps.
(5) Time and resources (money and equipment) available for the work.
(6) The availability and skill of personnel.
(7) A designer's knowledge of statistics and sampling theory.

Sampling designs There are two main types of sampling designs: random or probability, and non-random or systematic. Random sampling implies that each sampling unit from the total population of sampling units in the entire forest area has an equal chance of being selected. Forest inventories using random sampling require the use of aerial photographs or maps to establish a frame from which to draw sampling units using for example, random numbers. Where the forest is heterogeneous, it is recommended to stratify it into homogeneous subdivisions and random samples drawn from each of the subdivisions. The number of samples drawn from each subdivision is determined by the proportional importance of each subdivision.

This sampling design is referred to as stratified random sampling. For a given sampling intensity, stratification often yields more precise estimates of forest parameters than does a single random sample of the same size. Separate estimates of means and variances can be made for each of the forest strata.

The process of stratification can be further elaborated by the use of multistage sampling. In multistage sampling the forest is divided into a hierarchy of sampling units: primary, secondary, tertiary etc. The main advantage of this approach is the concentration of measurement work close to the locations of the chosen primary sampling units which would otherwise be spread over the entire forest area. Concentration of sampling units in fewer locations also reduces the risk of non-sampling errors. Furthermore, two-stage sampling often yields estimates of a required precision at lower cost than mono-stage sampling.

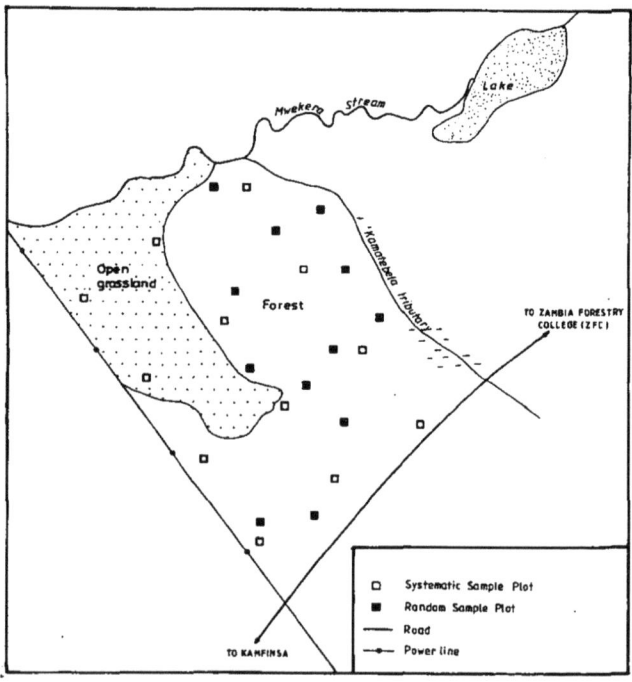

Figure 4.2 *Sample plot location at Mwekera*

In non-random or systematic sampling, the units which make up the sample are not chosen by the laws of chance but by personal judgement or systematically. Usually, but not always, systematic sampling units are spaced at fixed intervals throughout the inventory area or population. Because the sampling units are spread over the entire area, they usually give good estimates of population means and totals. This sampling design is often faster and cheaper to execute than random designs because the selection of sampling units is mechanical and uniform (see Box 4.1).

Size and shape of sampling units The size of the most efficient sampling unit depends on the variability of the forest. Larger sampling units are recommended in forests with variable composition, because small sampling units give high coefficients of variation. Small sampling units are also liable to bias unless care is taken in their siting and demarcation. However, in less variable forests small sampling units tend to increase the precision of estimates because for a given sampling intensity the number of independent small sampling units is larger than for larger units.

The ultimate choice of the size and shape of a sampling unit must be based on cost and precision considerations. The relative efficiency (RE) of different sampling plot sizes or shapes can be calculated as follows:

$$RE = \frac{SE_1^2 \, t_1}{SE_2^2 \, t_2} \qquad \text{Sampling plot selection (2)}$$

where
SE_1 = standard error (in %) for the base plot size or shape used in the comparison.
SE_2 = standard error (in %) for the other plot size or shape.
t_1 = cost of or time for establishing and inventorying the base plot size or shape.
t_2 = cost of or time for establishing and inventorying the other plot size or shape.

If RE is <1.0 then the base plot is more efficient, and when RE >1.0 then the other plot is more efficient. In some inventories, long narrow parallel transects may be more effective units as long as great care is taken in their demarcation and survey to avoid a low area to perimeter ratio, which tends to increase bias. Circular plots are easy to demarcate in open stands of trees with little ground cover or shrubs, but can be difficult to establish where the understorey is thick.

By plotting average sample plot data against increasing sample area, it is possible to determine graphically the minimum sample area. In theory this should reduce the variance about the average. However, in miombo, increasing sample area tends to reduce bias in the average estimate without necessarily reducing the variance (Figure 4.3), because of the great spatial variation in stocking (Chidumayo 1991b).

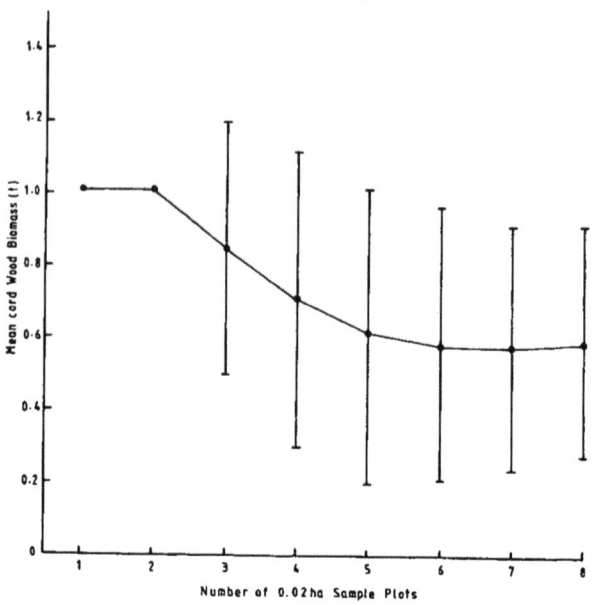

Figure 4.3 *Effect of increasing sampling area on biomass estimate (dots) and standard deviation (vertical bars) in dry miombo*

Past inventories of miombo The majority of sample plots used in miombo woodland inventories have been systematically selected and located (Trapnell 1959; Lees 1962; Lawton 1978; Chidumayo 1987a). Lawton (1978) used roads and tracks to locate traverses in each forest type delineated on aerial photographs oriented along a topographical sequence, extending from one watercourse through the interfluve crest (see Figure 1.4) to the next watercourse. Sample plots were systematically located at variable intervals along each traverse. Lees (1962) used inventory plots that were laid out systematically across contours at right angles from accurately measured and semi-permanently beaconed base lines.

The procedure for locating random plots is complex. Shackleton (1993) in a South African savanna used a co-ordinate system laid over aerial photographs and randomly generated sample sites which were marked on the overlay (see also Box 4.1). Each site in the field was measured from the closest identifiable point to ensure that samples were sited according to the distribution on the overlay.

A variety of plot sizes have been used (Figure 4.4), but the common sample plot size used in miombo woodland inventories is 0.4 ha (Lees 1962; Edmonds 1964; Chidumayo 1987a; Trapnell 1959). However, 20 × 20 m (Lawton 1978; Araki 1992; Stromgaard 1985b,1985c) and 40 × 40 m (Scholes 1990) plots have also been used. Fewer inventories have been based on circular plots (e.g. Zimba 1991). Large transects are used in studies of

80

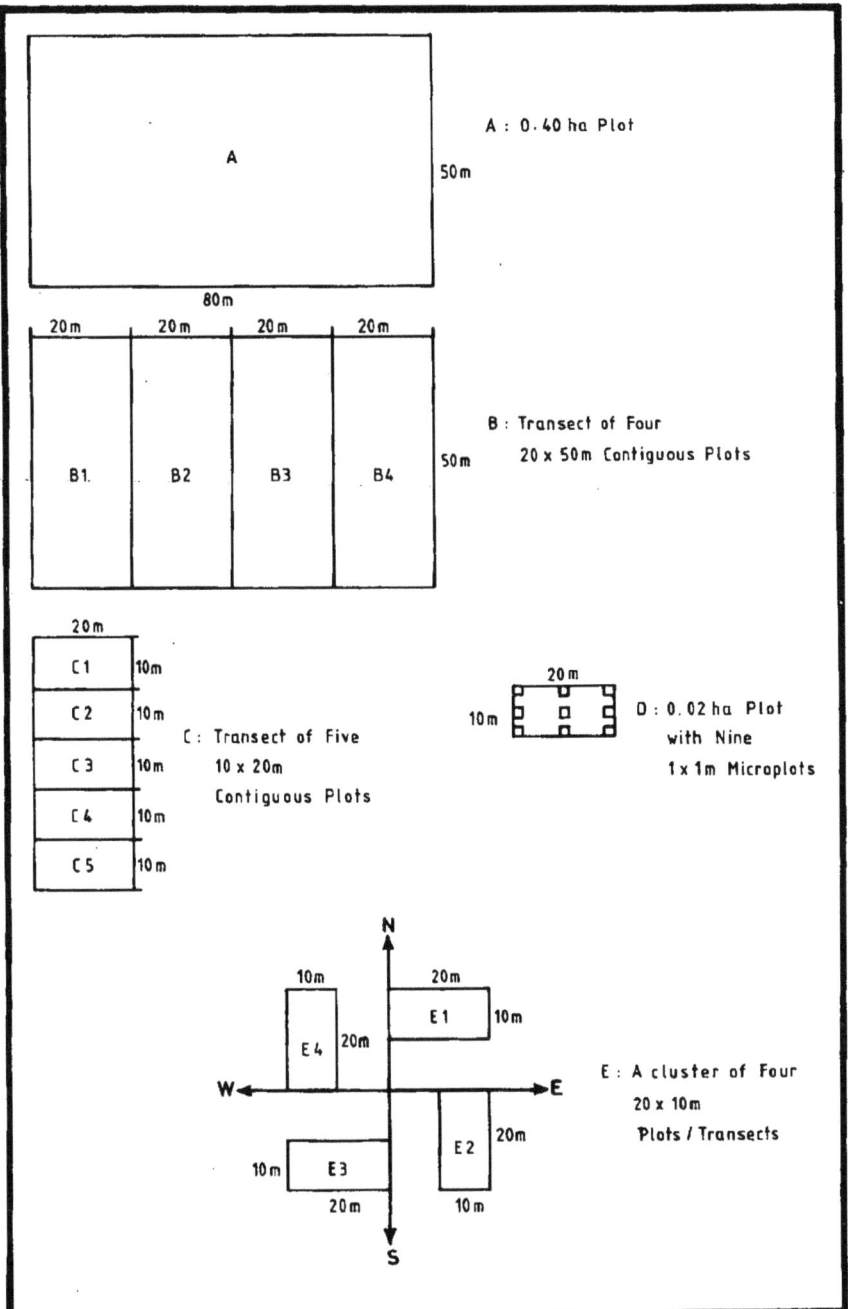

Figure 4.4 *Types of woodland inventory plots*

Box 4.2 *Assessing litter biomass in Mwekera coppice miombo plots using microplots.*

During the workshop to test the draft version of this handbook, in April 1995 a survey was made of surface litter biomass in Mwekera coppice miombo plots (each 0.4 ha) established in 1960 and maintained under fire protection, early burning and late burning regimes using systematically located 0.5 × 0.5 m microplots. Unfortunately the fire-protected plot was accidentally burnt during August/September 1994. In each plot six microplots were established along a transect bisecting the plot with an inter-microplot distance of 8.0 m. Where there was an impenetrable obstacle, such as an anthill thicket, along the transect the location of the microplot was shifted to facilitate sampling. Litter was collected from each microplot and separated into leaves (mainly tree leaves), pod valves, wood, grass and bamboo, and each was immediately weighed. In each plot a subsample of each litter type present was collected and weighed before oven-drying for 24 h at 80°C. The subsamples were then re-weighed to determine the oven-dry/field weight ratios. These ratios were 0.8 for leaf, pod valve and wood and 0.9 for grass and bamboo litter. The structure of the surface litter in the three fire treatment coppice plots was as shown in the table below (where PP is protected plot, EBP is early burnt plot and LBP is late burnt plot).

Microplot	Litter biomass (g/ 0.25 m^2)								
	Grass			*Leaves*			*Pod valves+wood+bamboo*		
	PP	EBP	LBP	PP	EBP	LBP	PP	EBP	LBP
1	0.0	1.8	0.0	14.4	86.4	4.8	62.4	12.8	3.6
2	0.0	0.0	3.6	6.4	182.4	4.8	19.8	72.0	0.0
3	0.9	0.0	0.0	9.6	58.4	3.2	76.8	723.0	3.6
4	0.0	16.2	0.0	9.6	25.6	6.4	18.0	523.7	17.6
5	9.0	16.2	0.0	35.2	30.4	16.0	97.6	70.4	8.0
6	0.0	0.0	3.6	9.6	83.2	4.8	97.6	63.2	12.8
Mean	1.7	5.7	1.2	14.1	77.7	6.7	62.0	244.2	7.6
SD	3.6	8.2	1.9	10.6	57.3	4.7	36.0	301.2	6.6

The only significant difference among the plots was in respect of leaf litter ($F = 8.04$, $P < 0.01$) with the Least Significant Range (LSR) of 23.9 g. The LSR value shows that only the early burnt plot had significantly more leaf litter than the other two plots which statistically had similar leaf litter biomass. The lack of significant differences in grass, pod valve, wood and bamboo litter among the plots was probably caused by the large spatial variation in the distribution of these litter types. Increasing the size of microplots would probably have reduced the variation. Contrary to expectation, the fire-protected plot did not have more litter than the early burnt plot, perhaps because the accidental August/September 1994 fire destroyed most of the litter biomass. It is also apparent from the results that decomposition does not remove all the litter at the end of the rainy season because litter from the previous year was still present in April 1995. In fact bamboo litter was caused by a widespread die-off of bamboo that occurred in the Copperbelt area during 1989/90, and this litter was still abundant almost five years later.

woodland-wildlife interactions. Guy (1981) used plots that were 5–100 m wide and 50–100 m long in Sengwa wildlife area in Zimbabwe. However, in studies of small-scale vegetation patterns, many smaller contiguous plots have been used. Chidumayo (1991b) used 10 × 20 m and 20 × 50 m contiguous plots while Campbell et al. (1988) used 10 × 10 m contiguous plots. Shea et al., (1993) used a cluster of transects each 15 m long and 10 m wide. Micro-plots (1–10 m^2) within larger tree census plots, are used to inventory seed banks, tree seedlings, litter, grasses and other herbs (Chidumayo 1993a; Ellenbroek 1987). Even smaller micro-plots (50 × 50 cm) have been used for such studies (Hood 1972; Shea et al., 1993) (also see Box 4.2).

If the census in a sample plot is to be repeated at a later date (as is the case with recurrent inventories), permanent boundary markers, such as beacons and posts, must be installed at the site and the location described and drawn on a field map to facilitate the future identification of the plot.

Types of data

Shrubs and trees During a forest inventory, each tree rooted in the sample plot is identified, counted and measured and the data recorded on a prescribed form. Trees may be identified fully by species name (e.g. *Brachystegia boehmii*), or incompletely by genus name (*Brachystegia* sp.). Unidentified trees can be recorded by their local names or simply entered as unidentified species. However, it is important to distinguish the different incompletely and unidentified trees by a number or letter (e.g. *Brachystegia* sp.1, *Brachystegia* sp.2 etc.) and parts of the tree (e.g. a branch with leaves, flowers or fruit) taken for subsequent identification by a taxonomist.

Trees in a sample plot are counted consecutively and marked to avoid double counting. In the case of repeat censuses that involve the precise location and identification of individuals, trees must be marked or tagged and given clear permanent numbers and the measurement position shown with a permanent mark during the initial census. Metal tags nailed to the trees are very useful for this purpose. Stems on multi-stemmed trees, which are common in regrowth miombo, should be distinguished (Figure 4.5).

Tree measurements include girth/diameter, height and crown diameter. In miombo woodland girth/diameter is measured at 1.3 m (breast height, bh) above ground (AG). But if the tree is branched or swollen at bh, measurement can be made immediately above or below the abnormality. Other workers have made measurements at 0.3 m (knee height) AG (Chidumayo and Chidumayo 1984) and 5–10 cm (ankle height) AG (Tietema 1993). However, measurements at stem base or just above the basal swelling are most appropriate for young coppice stools and saplings under 2 m tall.

These measurement positions are shown in Figure 4.5. For coppice and sapling stems, it is advisable to measure the diameter with calipers rather than the girth with a tape. Single measurements cannot be used to

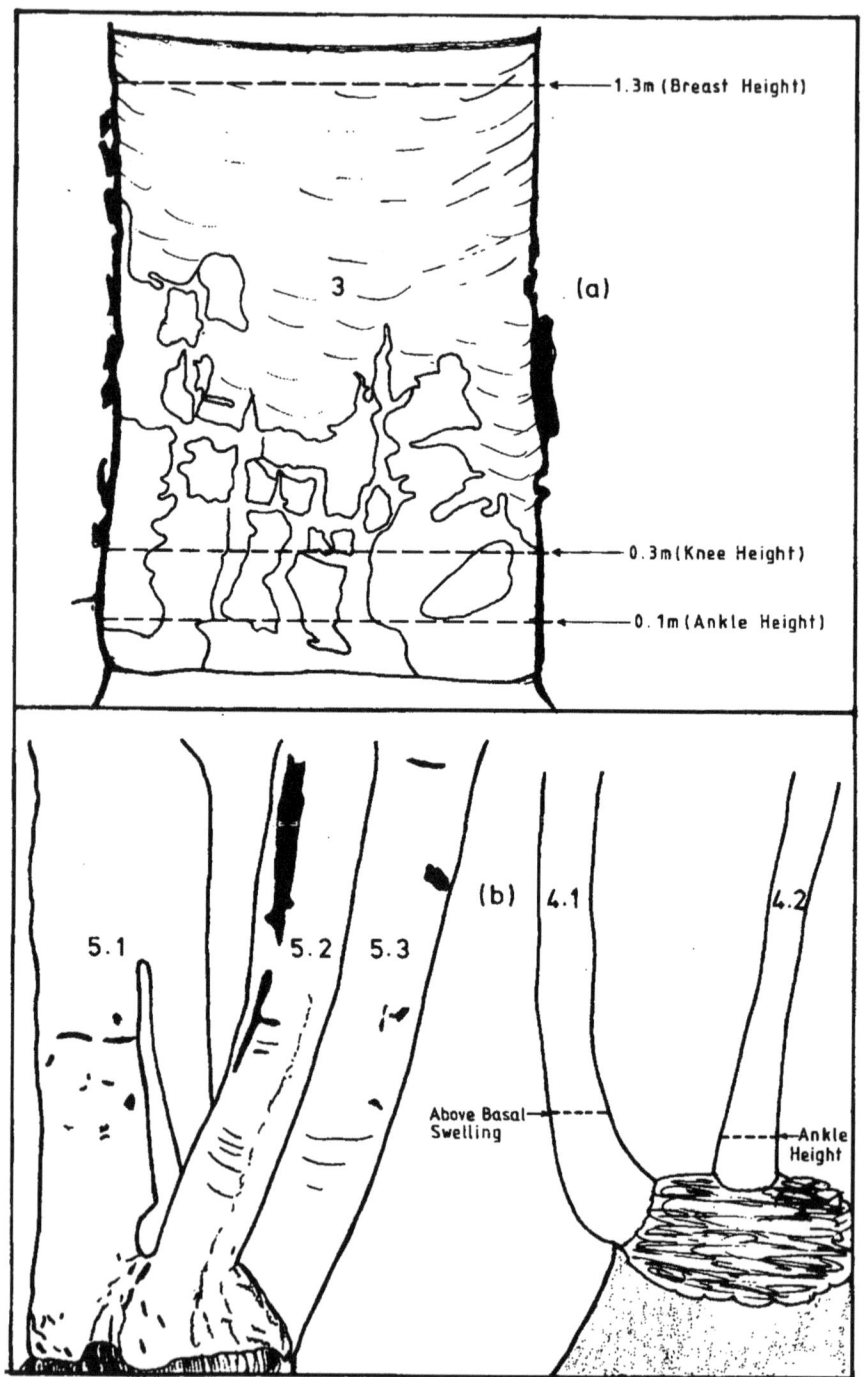

Figure 4.5 *Stem measurement positions and marking of stem*

determine temporal changes in stem growth. To determine age-specific changes in stem growth, repeated measurements of the same trees at different ages are necessary. Such studies have shown annual fluctuations in the stem girth increments of miombo trees (Chidumayo 1993a).

Crown diameter is measured along North–South and East–West axes and the average calculated (Tietema 1993). Tree height may be reliably estimated by experience; otherwise it can be measured by a variety of instruments, such as a hypsometer.

Other useful tree data that can be recorded during an inventory include phenology (leaf flush and shedding, leaflessness, flowering and fruiting) and tree condition (defoliated, fire scarred, stumped, pollarded, browsed, debarked, etc.).

Suffrutices and herbs The census of woody suffrutices and some herbs is more difficult than that of trees because such plants tend to have a colonial habit which makes the recognition of individuals extremely difficult. But density estimates can be made by counting shoots in micro-plots and expressing density as a number m^{-2}.

Soils It is common practice during a forest inventory to collect soil data. Such data are used in site description and assessments of site quality and potential. Edmonds (1964) collected data on a number of soil variables at his sample plots. These included soil colour, texture, compaction, depth to impediments, water table and humus content. Lawton (1978) collected 10–20 cm auger soil samples from 1.8 m deep soil pits at sample plots. For more detailed studies, soil samples are analysed for nutrient content, pH, bulk density and soil moisture (Stromgaard 1984; Chidumayo 1993a, 1993b). A brief description of soil analyses are presented below, but for more detailed coverage the reader is referred to Moore and Chapman (1976), Anderson and Ingram (1989) and other specialized textbooks in soil science.

Soil texture Soil texture is determined by particle-size distribution. Three diameter size ranges are used: sand (2–0.05 mm), silt (0.05–0.002 mm) and clay (< 0.002 mm). Since only the mineral material of diameter < 2 mm is used, the soil sample is first passed through a 2 mm sieve before removing aggregate binding agents, such as humus and iron, by chemical and/or physical processes. The particle-size separation is done by pipette, successive sedimentation and hydrometer methods (Day 1965). On the basis of the particle-size composition, the soil texture class can be determined from a textural triangle (Figure 4.6).

Organic matter Soil organic matter has a complex composition and there is no standardized method of determining its content in the soil. However, because carbon makes up about 50 per cent of the total organic matter, indirect estimates of organic matter are based on carbon determination. This is done by either quantitative combustion of carbon to carbon dioxide

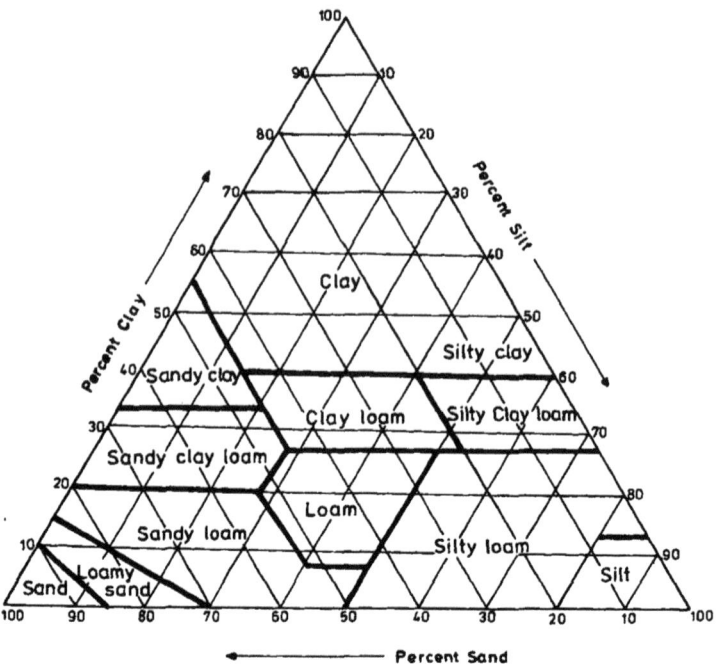

Figure 4.6 *Textural triangle for determining soil texture classes*

or the Walkley and Black method (Walkley 1946), which is based on chromate reduction. The Walkley and Black method is more commonly used because it is less elaborate and requires less expensive equipment than the carbon combustion method. After the soil sample (for example, 1 g air-dry soil) has been oxidized in an excess of an oxidizing solution, the excess chromate ions are back-titrated in a standard reduction solution.

Soil moisture and bulk density Soil moisture content is the mass loss in a soil sample during oven-drying at 105 °C for 24 h. It is expressed as percentage of oven-dry soil mass. The soil sample is obtained by a soil auger and placed in a sealed tin and weighed before and after oven-drying to determine loss of mass.

Bulk density is the mass of dry soil per unit volume, and is expressed as g cm^{-3}. The volume includes the enclosed pore space in between soil particles. A simple method of determining soil bulk density is to collect soil cores or monoliths of known volume and oven-drying these at 105 °C for 24 h. Soil bulk density is then calculated as:

$$\text{Bulk density} = \frac{\text{Soil mass}}{\text{Soil volume}} \qquad \text{Soil bulk density} \quad (3)$$

Bulk density can be used to convert water percentage by weight to water percentage by volume and for calculating total pore space. It can also be used to estimate water saturation point (a measure of water-holding capacity) by the following equation (Solbrig 1991):

$$\text{Water saturation point} = \frac{\text{Bulk density}}{2.64} \quad \text{Water saturation point (4)}$$

Soil pH Soil pH measures the concentration of hydrogen ions in the soil solution and is referred to as active acidity. This contrasts with exchangeable acidity which measures the amount of hydrogen and aluminium ions adsorbed to soil colloids. On the basis of pH, a soil can be classified as acid (pH < 7), neutral (pH = 7) or basic (pH > 7). The pH of a soil suspension is measured by a glass-calomel electrode-pH meter. Neutral salts, such as potassium chloride and calcium chloride are added to the suspension to liberate hydrogen and aluminium ions. In 0.01 M calcium chloride, the soil solution pH is reduced by 0.5 units, while in potassium chloride the pH is reduced by 1–1.5 units compared to values in salt-free water suspensions.

Soil nitrogen (N) The total nitrogen content in the soil is used to estimate the carbon:nitrogen ratio, as total N content bears little relationship to mineral N availability in the soil for plant uptake. Total N is analysed by the classical Kjeldahl method (Bremer 1960). This method is based on the destruction of organic matter in which N occurs as part of amino-acids. The organic matter is decomposed by sulphuric acid and the amino-N liberated as ammonium-N. This is then captured in excess boric acid and the solution titrated with 0.01 M hydrochloric acid.

Extractable phosphorus (P) Extractable or available P is usually determined by the Bray 1 method (Bray and Kurtz 1945) using air-dry soil in an extraction solution of 1.0 M ammonium fluoride, 0.5 M hydrochloric acid and distilled water.

Exchangeable bases To extract exchangeable cations, an excess of ammonium ions are added to an air-dry soil sample to replace the adsorbed exchangeable cations. The suspension is filtered through a Watman filter paper, after shaking for 30 minutes to equilibrate. The concentration of potassium and sodium cations in the filtrate is measured using a flame photometer while the concentration of calcium and magnesium is measured by the atomic absorption spectrophotometer (Hesse 1971).

Site history
Other habitat data recorded during a forest inventory may relate to site history, such as evidence of previous disturbances: cultivation, burning, felling, tree damage by herbivores or frost, charcoal-making etc. Such additional data improve the interpretation of inventory data.

Assessments

Plant biomass

Shrubs and trees The aim of the majority of miombo woodland inventories has been to estimate basal area and solid volume (SV) of timber and pole trees (> 15 cm girth at bh, or gbh) and stacked volume (STV) of cord wood (Edmonds 1964; Lees 1962). Only recently have biomass estimates been made (Araki 1992; Chidumayo 1990a, 1991b, 1993a, 1993b; Stromgaard 1985b, 1985c). Whereas the primary data for volume and biomass estimates are derived from destructive samples of felled trees (Appendix 2), basal area estimates are derived from girth or diameter measurements of non-destructive samples of standing trees. The data from felled trees are then used to develop regression equations for estimating volume and biomass from girth or diameter measurements of standing trees, either alone or in combination with height and crown diameter data (Box 4.3).

$$BABH = \frac{G^2}{12.56} = \frac{D^2 \times \pi}{4} \qquad \text{Basal area (5)}$$

where D is diameter and G is girth at bh; $\pi = 3.143$.

The basal areas of individual trees in a sample plot are pooled, and the sum converted by an appropriate factor, depending on the plot size, to give m^2 ha^{-1}.

Basal area may also be calculated for different stem size classes and multiplied by stem frequency in each class, and the products summed over all the classes. Chidumayo and Chidumayo (1984) used stem size classes and estimated BABH in dry and wet miombo at 11.4 m^2 and 14.2 m^2 ha^{-1}, respectively. Using the same method, Endean (1968) found a BABH of 16.5 m^2 ha^{-1} in undisturbed wet miombo while Malaisse (1984) found a BABH of 33.0 m^2 ha^{-1} in a dry evergreen miombo forest in southern Zaire, by summing basal areas of individual trees in the sample plot. See also Box 4.1.

Solid volume (SV): Solid volume (SV) is expressed in cubic meters (m^3) ha^{-1} and calculated using the following formula:

$$SV = \left[\frac{D^2 \times \pi}{4} \right] \times L \qquad \text{Solid volume (6)}$$

where L is the length of the bole/pole and D is the diameter at mid-length or the average of bottom and top diameters.

Diameter is measured over or under bark. Volumes of individual stems in a sample plot are pooled and converted to SV (m^3) ha^{-1} by an appropriate factor, depending on plot size. From sample data of felled stems, regression

Box 4.3 *Developing biomass equations using correlation and regression analysis.*

Diameter at breast (1.3 m above ground) height (DBH), total height (H) and biomass were measured on a sample of felled *Julbernardia globiflora* trees and after the biomass was converted from fresh to oven-dry the data were summarized as follows:

DBH (cm)	H (m)	Oven-dry biomass (kg)		
		Cordwood	Brushwood	Leaf
2	2.5	0.5	2.0	0.4
3	3.4	0.9	2.0	0.4
4	4.3	2.7	3.3	0.6
5	4.6	4.1	4.0	0.7
6	5.8	6.6	4.8	0.9
7	5.7	9.2	6.7	1.3
9	6.4	16.2	8.5	1.7
11	7.4	24.3	9.0	1.2
13	8.8	53.2	20.0	4.2
14	8.5	66.3	23.0	4.1
15	8.7	71.7	18.0	2.8
20	9.4	141.6	24.0	4.1
25	10.8	243.0	38.0	8.4

From these data it is possible to develop regression equations for estimating biomass of standing *J.globiflora* trees from DBH and H as independent variables which are easier to measure on a tree than biomass. However, because of the very large range in biomass between the smallest and largest tree, it is advisable to develop separate equations for small (<10 cm dbh) and large (> 10 cm dbh) trees. In some cases logarithmic transformation of the data can overcome this problem. The analysis starts with the calculation of total correlation between each biomass component and the two independent variables (dbh and h) for each tree size class. This analysis gave the following results:

Biomass component	Total correlation coefficient (r)			
	With DBH		With H	
	Small trees	Large trees	Small trees	Large trees
Leaf	0.97	0.87	0.89	0.95
Brushwood	0.98	0.91	0.91	0.97
Cordwood	0.97	0.99	0.89	0.96

Since both dbh and h gave very high and significant correlation with biomass components, it is necessary to calculate partial correlation (i.e., correlation with one independent variable while the other variable is held constant) of each biomass component with each of the two independent variables.
The results of this analysis are shown in Box 4.4.

Box 4.4 Results of biomass equations using correlation and regression analysis (Continued from Box 4.3)

Biomass component	Partial correlation coefficient (r)			
	with DBH		with H	
	Small trees	Large trees	Small trees	Large trees
Leaf	0.38	0.48	−0.05	0.81
Brushwood	0.25	0.59	−0.04	0.78
Cordwood	0.94	0.51	−0.80	0.28

Partial correlation analysis revealed that there was little correlation between H and leaf and brushwood biomass among small *J.globiflora* trees. The regression equations for estimating biomass can therefore be developed on the basis of DBH and H. In order to develop biomass equations regression coefficients for the independent variable(s) and the corresponding constant(s) should be calculated using regression analysis. This procedure resulted in the following equations for estimating each biomass component for each *J. globiflora* tree size class:

Biomass component (kg)/ tree size class		Regression equation
Leaf:	small trees	0.20dbh − 0.15
	large trees	0.17dbh + 2.78h − 17.92
Brushwood:	small trees	0.99dbh − 0.61
	large trees	−0.17dbh + 8.91h − 54.79
Cordwood:	small trees	3.79dbh − 2.76h + 0.86
	large trees	13.67dbh + 7.84h − 193.22

These equations, when applied to the observed data for independent variables, give the following biomass estimates compared to the observed values:

DBH (cm)	H (m)	Biomass (kg)					
		Cordwood		Brushwood		Leaf	
		Observed	Estimated	Observed	Estimated	Observed	Estimated
2	2.5	0.5	1.5	2.0	1.4	0.4	0.3
3	3.4	0.9	2.8	2.0	2.4	0.4	0.5
4	4.3	2.7	4.1	3.3	3.4	0.6	0.7
5	4.6	4.1	6.9	4.0	4.3	0.7	0.9
6	5.8	6.6	7.4	4.8	5.3	0.9	1.1
7	5.7	9.2	11.4	6.7	6.3	1.3	1.3
9	6.4	16.2	16.9	8.5	8.3	1.7	1.7
11	7.4	24.3	15.2	9.0	9.3	1.2	0.8
13	8.8	53.2	53.5	20.0	21.4	4.2	4.3
14	8.5	66.3	64.8	23.0	18.6	4.1	3.3
15	8.7	71.7	80.0	18.0	20.2	2.8	3.7
20	9.4	141.6	153.9	24.0	25.6	4.1	4.8
25	10.8	243.0	233.2	38.0	37.2	8.4	7.9

equations can be developed for estimating SV from D and L measurements. Such equations take the form of:

$$SV = a + bD^2L \qquad \text{Solid volume regression (7)}$$

where a and b are regression coefficients.

As can be appreciated, only SV of clear boles and straight branch poles can be assessed. Obviously, this limits the use of SV as a measure of total wood in a forest stand, especially in tropical forests where trees are characterized by a large proportion of branch wood (see Appendix 2). Lees (1962) estimated SV of the exploitable growing stock (> 45 cm gbh and > 2.4 m bole height) in wet miombo using girth classes. The inventory did not involve tree felling, and volume estimates were derived from volume tables in the Forest Department prepared from regression equations. A factor of 0.75 was used to convert over-bark to under-bark volume. The exploitable growing stock was estimated at 1.7 m^3 under bark ha^{-1}.

Edmonds (1964) assessed SV in dry miombo in central Zambia on felled 0.4-ha plots. The assessment included boles (> 15 cm gbh and > 1.5 m length) and poles from branches (> 1.5 m length). The measurement data used in the analysis included total length; bottom, mid-length and top girth over bark; and bottom and top bark thickness. From these data, SV over bark of boles and poles was estimated at 7.45 m^3 ha^{-1}. Edmonds also measured the stacked volume (STV) of non-bole/pole cord wood and visually estimated head loads of non-cord wood firewood. The average STV of cord wood was 103 m^3 ha^{-1}.

Biomass. Basal area and volume measurements cannot be used to determine total woody plant biomass, which is mainly made up of roots, AG wood and leaves. This weakness is overcome by biomass (B) measurements. Biomass is expressed as oven-dry or green (fresh) mass (t) ha^{-1}. Biomass equations or tables can be developed from sample felled trees data. But whereas only one volume equation for each species is required to estimate stem SV, a set of biomass equations is required for each species or species group to estimate all biomass components: stem wood, cord wood, twig or brush wood, leaves and roots. The equations are normally based on girth or diameter at bh as the independent variable, although tree height (H) and crown diameter may also be included (Box 4.3). The equations may take the following forms:

$$B_i = a_i + b_iD \qquad \text{Biomass equation (based on girth) (8)}$$

$$B_i = a_i + b_iD^2H \qquad \text{Biomass equation (based on height and girth) (9)}$$

where B_i is the biomass of tree component i (i.e. stem wood, cord wood, twigs, leaves etc.), D is tree diameter, H is tree height and a and b are regression coefficients.

Biomass equations for some miombo woodland trees are given in Appendices 3–5 and summarized in Table 4.2.

To obtain data for developing biomass equations, trees of different sizes are cut, and all or subsamples of biomass components are weighed immediately after felling to avoid errors caused by moisture loss. Fresh weight data are converted to dry weight by oven-dry/fresh weight ratios (Appendix 6) found after drying biomass samples at 80 °C for 48 h and 72 h for leaves and wood, respectively (Chidumayo 1993a, 1993b). The loss in weight is

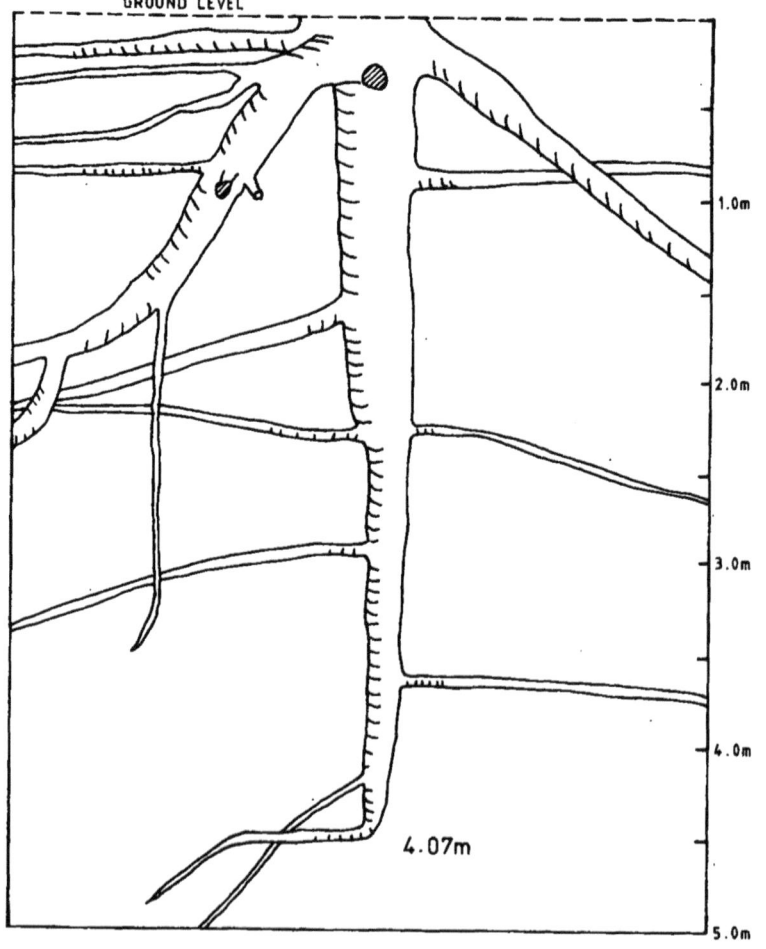

Figure 4.7 *Root structure in Brachystegia spiciformis (Based on Savory 1962)*

Table 4.2 *Generalized biomass equations for estimating above ground woody plant biomass (kg, oven dry) of small (< 10 cm diameter at breast height, dbh) and large (>10 cm dbh) stem in miombo woodland*

Biomass component	Stem class	Equation	R-squared
Leaves	Small	0.27 dbh – 0.47	0.92
	Large	0.59 dbh – 5.13	0.94
Brush wood	Small	0.51 dbh – 0.57	0.92
	Large	2.00 dbh – 9.40	0.85
Cord wood	Small	2.23 dbh – 6.44	0.92
	Large	17.43 dbh – 188.84	0.94

Based on Chidumayo (1993a and 1993b)

equated to biomass moisture content (MC). Average wood MC (as a percentage dry weight) of miombo trees is 77 per cent (Table 3.4) while that of mature green leaves is 110 per cent. Because miombo trees are deciduous, the assessment of leaf biomass should be done after the growing season but before the onset of the peak leaf-fall period. This period is from February to July (Chidumayo 1993a). Leaf production can be estimated by monitoring tree leaf fall in permanent quadrats throughout the year (see Figure 3.2) or by destructive samples obtained after the end of the shoot extension period but before the onset of the peak leaf fall period. These destructive samples are used to develop leaf biomass equations using girth or diameter data (Box 4.3, Table 4.2.)

Leaf production estimates in dry miombo using the two methods are compared in Table 4.3. In old-growth miombo the methods give close estimates but differences increase with decreasing regrowth age. Generally, biomass equations give lower leaf biomass estimates than the litter-fall method, because equations exclude or underestimate the leaf biomass of small trees and saplings.

Sampling tree root biomass is more difficult than sampling AG biomass. Miombo trees have very deep roots (Figure 4.7). The average tap root depth of canopy and understorey trees is 2.4 m and 1.5 m, respectively, and lateral roots which are diffused at different depths extend for 4–15 m from root crowns (Savory 1962). Excavating and recovering roots in miombo woodland is therefore very difficult and laborious.

Chidumayo (1993a, 1993b) used two types of samples to estimate woody plant root biomass in dry miombo (Figure 4.8). The root biomass was weighed of a particular tree within a 1.0 m radius of its root crown, up to a depth of 1.0 m, in felled sample plots. Roots of other woody plants in this sample zone were excluded. The pooled data for all individual trees gave an estimate of total tree-specific root biomass of 1.3 kg m^{-2}. Another set of root samples was obtained in 1 × 1 m and 1.5 m deep soil pits located on the edge of each sample plot (see Figure 4.8). These deep pits were at least

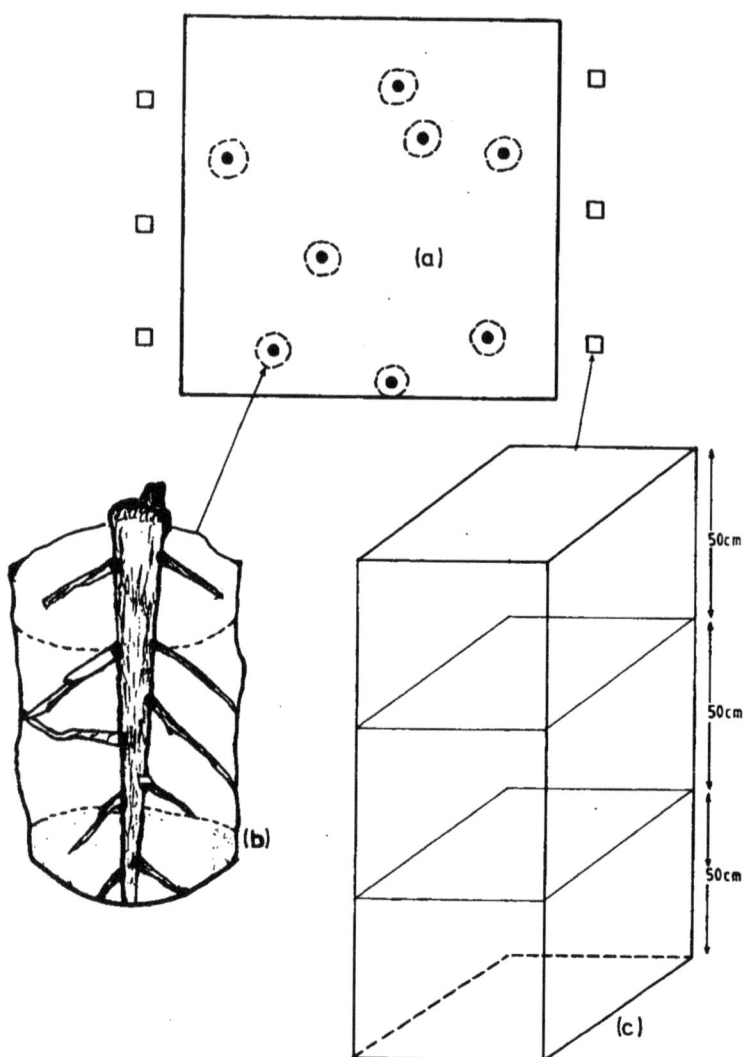

Figure 4.8 *Types of soil pits for assessing root biomass in miombo. (a) cylindrical pits around each tree for tree-specific roots (b). (c) = prismatic pits for diffused lateral roots.*

> 1.0 m from any tree to avoid re-sampling tree-specific root biomass. All the roots in each soil pit were weighed separately for each 50 cm-thick soil layer, after separating them into woody plant, grass and other herb categories. These root biomass samples were used to estimate the diffused lateral root biomass and were expressed as kg m^{-2}. In old-growth miombo this averaged at 2.0 kg m^{-2}. Dry/fresh weight ratios were determined from

Table 4.3 *Estimates of annual tree leaf production (oven dry mass) in central Zambia dry miombo during 1991–1993*

	Mean annual leaf production (g m^{-2})	
Woodland type	Biomass equation method	Leaf fall method
Old-growth	230	249
19–21 yr old regrowth	201	248
9–11 yr old regrowth	222	297

Based on Chidumayo (1993a)

oven-dried samples at 80 °C for 72 h to convert fresh to dry weight. The moisture content in roots in miombo trees is similar to AG stem wood (Chidumayo unpublished). The data from the two types of root samples were pooled to give an estimate of 3.3 kg m^{-2} or 33 t ha^{-1} for total root biomass at the old-growth dry miombo sample sites in central Zambia (Chidumayo 1993a, 1993b).

This is obviously an underestimate of root biomass in miombo, because fine roots were not included and sample soil pits were shallower than the rooting depths of miombo trees (see Figure 4.7). However, most samples of forest root biomass based on soil pits exclude tap roots and root crowns (Golley *et al.* 1975), which was not the case in the study by Chidumayo (1993a, 1993b). The distribution of lateral root biomass in the deep pits was 76 per cent in the 0–50 cm, 21 per cent in the 51–100 cm and 3 per cent in 101–150 cm soil depths. This pattern of root biomass distribution suggests that the large roots were concentrated in the 0–100 cm depth. Other studies have shown that most of the large root biomass is found within 0–60 cm soil depth (Greenland and Kowal 1960; Okali *et al.* 1973; Rutherford 1983).

Relationships between biomass variables. Since basal area is easy to calculate from girth or diameter measurements, simple regression equations can be developed to estimate log or cord wood SV and biomass. These equations are given in Table 4.4. Small stems (< 10 cm dbh) have been excluded in Table 4.4 because these show a very weak correlation between basal area and SV or the biomass of cord wood. Because cord wood for fuel is normally measured in stacked volume (STV), it is useful to develop equations for estimating biomass from measurements of systematically piled cord wood (Table 4.4). From the equation for large stacks, it is estimated that 1 m^3 of cord wood from miombo trees has a dry mass of 306 kg, which is similar to an estimate in Chidumayo (1991b). Wood biomass characteristics of some miombo trees are given in Appendix 2.

Wood specific density is the ratio between biomass and SV and is expressed either as g cm^{-3} or kg m^{-3}. Unfortunately wood density can vary within a stem or species and between species (Table 4.5). Within-stem variation is mainly caused by heart wood rot, which creates a hollow stem and distorts the relationship between volume and wood mass.

Table 4.4 *Specific wood density (g cm⁻³) of some miombo woodland trees. Standard deviations are shown in brackets.*

Species	Specific density (g cm⁻³) of some miombo wood from different parts of the stem			
	Stem top	Mid-stem	Stem base	Whole stem
Brachystegia boehmii	0.96	0.63	0.65	0.75
Brachystegia spiciformis	0.82	0.75	0.63	0.74
Isoberlinia angolensis	0.54	0.55	0.60	0.57
Julbernardia globiflora	0.81	0.68	0.62	0.70
Uapaca kirkiana	0.47	0.50	0.45	0.47
Uapaca nitida	0.62	0.50	0.43	0.52

Based on Chidumayo (1993c)

Herbaceous plants Forest inventories in miombo rarely include quantitative assessment of non-woody plants. Miombo woodland is important for grazing (see Chapter 2) and the supply of thatching grass for buildings, especially in rural areas. Herbaceous plants also account for a significant proportion of the fuel for igniting and sustaining wild fires in miombo (see Chapter 3). The roots of some herbs are important sources of food and medicines. An assessment of the biomass of herbaceous plants should therefore be useful to miombo woodland management.

Herbaceous plant biomass can be sampled in micro-plots of 0.5 × 0.5 m or 1 × 1 m size. Above ground biomass is harvested by clipping plants close to the ground (< 5 cm AG). If the objective is to estimate peak biomass, sampling should be done at the end of the growing season, which in miombo woodland is from February to May (Figure 4.9). Below ground herbaceous biomass is assessed in soil pits dug at the clipped micro-plots, as described in the previous section. However, almost all the roots of herbaceous plants are found in the 0–50 cm soil depth (Chidumayo 1993a). The biomass can be separated into grass and other herbs, weighed and the results expressed in g m⁻². Dry/fresh weight ratios calculated from subsamples dried at 80 °C for 48 h are used to convert fresh to dry weight. The MC

Table 4.5 *Regression for estimating cord wood volume (SV) and biomass (oven dry weight) from basal area at breast height (babh) and biomass from stacked volume (STV) for different miombo woodland biomass classes.*

Independent	Dependent	Biomass class	Equation	R-squared
Basal area (m²)	SV (m²)	Large stems (>10 cm dbh)	0.0005 babh+0.021	0.74
Stacked volume (m³)	Biomass (kg)	Large stems (>10 cm dbh)	0.029 babh+28.98	0.73
	Biomass (t)	Stand	4.45 babh−0.11	0.86
	Biomass (kg)	Small stacks (<1 m³)	316 STV+4.0	0.75
	Biomass (kg)	Large stacks (>1 m³)	85 STV+21.0	0.99

Based on Chidumayo (unpublished)

Table 4.6 *Changes in moisture content (% of dry weight) of above ground green grass biomass in dry miombo woodland*

Month	Average moisture content (% dry weight)	Standard deviation
December	255	60
January	241	86
February	169	59
March	153	21
April	96	21
May	89	15

Based on Chidumayo (unpublished)

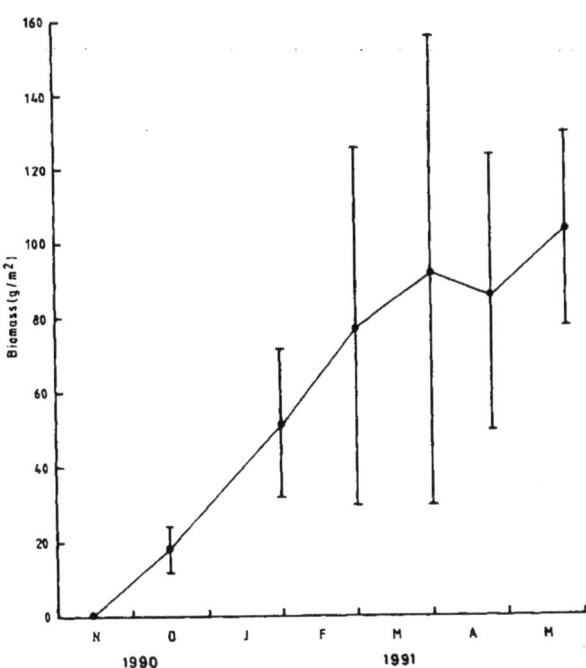

Figure 4.9 *Grass production in dry miombo*

of AG grass biomass in miombo woodland decreases steadily as the growing season progresses (Table 4.6).

Biomass chemical composition The nutrient content in plant biomass can be analysed by a number of methods. Phosphorus content can be determined by calorimetry using ascorbic acid, while potassium, calcium, magnesium and sodium can be determined by atomic absorption after ashing a gram of finely ground dried material in a muffle furnace for two hours at

Table 4.7 *Chemical composition of plant biomass in woodland in central Zambia*

Plant/Biomass type	Average chemical composition (% oven-dry biomass)				
	Ash	Carbon	Nitrogen	Moisture	Hydrogen and Oxygen
Woody plant:					
Green leaves	5.28	43.81	1.35	5.75	43.81
Stem wood	4.72	45.04	0.40	4.37	45.47
Roots	9.59	41.92	0.39	6.18	41.92
Grasses:					
Green stems + leaves	8.87	43.88	0.60	2.77	43.88
Roots	35.46	30.67	0.44	2.76	30.67

Based on Chidumayo (1994a % unpublished data)

450 °C and boiling in nitric acid. Nitrogen is determined by the micro Kjeldahl method. A gram of finely ground material is digested in sulphuric acid using a catalyst mixture and diluted with distilled water after cooling. Nitrogen is then determined by titration in 0.01 N hydrochloric acid after distillation in boric acid indicator solution.

Carbon content is more difficult to determine, but can be estimated by first removing volatile matter from the oven-dried sample at 1000 °C for seven minutes. The remaining material is cooled and weighed before destroying the carbon matter at 1000 °C for two hours, after which the ash residue is cooled and weighed. By calculating the weight loss at each stage of the process, the percentage volatile matter, carbon and ash can be determined. Other important plant chemicals that may be analysed include resins, proteins, carbohydrates and so on. Some chemical characteristics of plant biomass in miombo are given in Table 4.7.

Species diversity

The idea of species diversity contains two concepts: species richness and evenness. Species richness refers to the number of species in a community. Most communities of plants contain a few dominant species with numerous individuals and many rare species with very few individuals. Evenness measures the distribution of individuals among the different species in the community.

Species density (number/area) is the simplest measure of species diversity. Other measures involve the calculation of mathematical indices, and the Simpson Index of diversity (D_v), which measures the increase in the number of species as sample size increases (Krebs 1989), is the easiest of these indices. It is calculated in two steps. The dominance index (D) which measures the distribution of individuals among the species is calculated first by the formula:

$$D = \sum \left(\frac{n_i}{N}\right)^2 \qquad \text{Simpson's dominance index (10)}$$

where n_i is the number of individuals of species i in the sample and N is total individuals of all the species in the sample.

The D_v is then derived by subtraction as follows:

$$D_v = 1 - D \qquad \text{Simpson's diversity index} \quad (11)$$

It is apparent from (11) that if a sample has a high dominance index (D), it will have a low species diversity, and vice versa. Whereas species density requires only a census of the species in a sample plot, it is also necessary to quantify its relative importance (i.e., numbers, biomass, cover, productivity) in order to calculate D and D_v in (10) and (11).

Another commonly used measure of species diversity is the Shannon-Wiener index which is defined as:

$$H' = -p_i \log(p_i) \qquad \text{Shannon-Wiener diversity index} \quad (12)$$

where p_i is the same as n_i/N in equation (10).

The Shannon-Wiener index is strictly used for random samples, although in practice it has been used for non-random samples. The Shannon-Wiener index increases with the number of species in the community and in theory can reach very large values, but in practice does not exceed 5.0 for biological communities. The most commonly used index of evenness is based on the Shannon-Wiener diversity function and is defined as:

$$E = \frac{H'}{\log S} \qquad \text{Index of species evenness} \quad (13)$$

where E is the evenness and S is the total number of species in the sample.

Examples of calculated indices of species diversity are shown in Box 4.5 based on data in Appendix 7.

The species–area curve is a method of presenting species diversity in a graphical form. The curve shows the rate at which new species are encountered as the sample area or size is increased. The species–area data are obtained from contiguous sample plots (see Figure 4.4 and Appendix 8). The curve is constructed by plotting cumulative species on the y-axis against cumulative sample area or size on the x-axis (Figure 4.10) which confirms that miombo woodland has a higher species diversity than other forests (Table 1.3).

Although the species–area curve can be used to compare two communities, there are other similarity measures. Such measures are mainly descriptive coefficients with no easy measure of confidence intervals (Krebs 1989). The simplest of these are binary coefficients which are based on species presence and absence data for the two communities being compared. The binary data are presented in a 2 × 2 table as shown in Table 4.8.

Table 4.8 Two-by-two table for calculating species diversity

	Community/Sample A	
Community/sample B	Species present	Species absent
Species present	a = species present in samples A and B	b = species present in sample B but absent in sample A
Species absent	c = species present in sample A but absent in sample B	d = species absent in both samples A and B

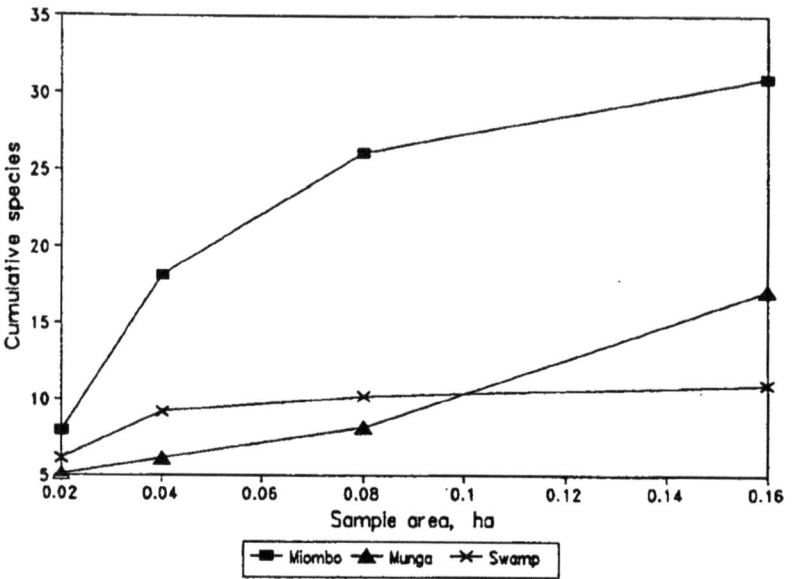

Figure 4.10 *Species-area curves for three forest types in central Zambia*

Usually d is not ecologically meaningful unless the flora of the community from which the samples have been drawn is well known. Four of the commonly used binary similarity coefficients are presented below. The coefficients range from 0 (no similarity) to 1 (complete similarity), but sample size and species richness may affect the maximum value (Krebs 1989). These coefficients are calculated in Box 4.4 using data from Appendix 7.

$$S_j = \frac{a}{a+b+c}$$ Jaccard's coefficient of similarity (14)

$$S_s = \frac{2a}{2a+b+c}$$ Sorensen's coefficient of similarity (15)

Box 4.5 *Calculation of indices of species diversity.*

Enumeration and estimated biomass data for three miombo woodland communities occupying three habitats on the edge of a dambo are given in Appendix 7. The species structure of the three communities are summarized below.

Sample (0.16 ha)	Total species (S)	Total individuals	Total above ground biomass (kg)
Creek zone	23	134	15 966
Wash zone	23	204	7 544
Scarp zone	24	317	7 116

Species diversity for the three miombo samples are calculated using the formulae given in the text: Simpson's dominance index, D (10), Simpson's diversity index, D_v (10); Shannon-Wiener diversity index, H' (12); Species evenness index, E (13). The calculated species diversity indices are summarized below:

	Measure of relative importance					
	Numbers			Biomass		
Index	Creek zone	Wash zone	Scarp zone	Creek zone	Wash zone	Scarp zone
D	0.084	0.095	0.130	0.152	0.158	0.177
D_v	0.916	0.905	0.870	0.848	0.842	0.823
H'	1.183	1.166	1.013	0.999	0.955	0.897
E	0.869	0.856	0.734	0.734	0.731	0.650

Note that although species richness was almost the same in the three samples, species diversity was somewhat different with the creek zone miombo having the highest diversity. Also note that although the wash zone and scarp zone samples had similar above ground biomass, species diversity based on biomass was different in the two samples, with the wash zone community having a higher species diversity.

$$S_{sm} = \frac{a+b}{a+b+c+d} \qquad \text{Simple matching coefficient of similarity (16)}$$

$$S_B = (\sqrt{ad}) + \frac{a}{a+b+c} + (\sqrt{ad}) \qquad \text{Baroni-Urbani and Buser coefficient of similarity (17)}$$

Table 4.9 *Percentage similarity for three miombo samples in Central Zambia*

Sample pair	Percentage similarity	
	Number of individuals	Above ground biomass
Creek vs Wash zone	26.2	22.8
Creek vs Scarp zone	54.0	40.7
Wash vs Scarp zone	23.4	20.9

One of the most useful quantitative similarity coefficients is the Percentage Similarity (P) which is defined as:

$P = $ minimum (pA_i, pB_i) Percentage coefficient of similarity (18)

where p_i is the percentage of species i in community A and B and summation is made of the minimum p_is for each A and B pair.

This measure of similarity is relatively little affected by sample size and species diversity. The calculated P for the three miombo samples on the edge of a dambo in central Zambia (see Appendix 7) is given Table 4.9. Note that data based on numbers of individuals consistently gave higher coefficients than biomass data. However, both data show greater similarity between creek zone and scarp zone miombo than elsewhere.

Very little is known about herbaceous plant species diversity in miombo woodland. The annual habit and great phenological diversity of herbaceous plants requires repeated censuses of sample plots in order to obtain a complete species inventory. In a wet miombo area in northern Zambia, herbaceous plants made up 65 per cent of the flora (Astle 1968–69) while Chidumayo (1993a) found that out of 184 species of flowering plants at four dry miombo sites in central Zambia, 73 per cent were herbaceous and woody suffrutices. These observations indicate that flora in miombo is dominated by herbaceous plants, although these are rarely enumerated.

Seeds and seedlings

Conducting repeated or annual censuses of seeds in permanent plots is perhaps the best way of studying changes in seed production in miombo. However, such censuses should be accompanied by observations on the flowering and fruiting of marked trees over a period of several or more years to assess the annual frequencies of fruit production. In addition seed production should also be assessed by counting mature fruits and their seeds on sample trees, and seeds should be separated into damaged/bad and good ones so that pre-dispersal seed damage can be determined. Once the number or frequency of fruiting trees, average fruits per tree and number of seeds per fruit are known, it is possible to estimate gross seed production in a given area (Table 4.10).

Box 4.6 *Calculation of similarity coefficients for binary data.*

For calculating binary similarity coefficients using data in Appendix 7, the total number of species in the whole dambo edge community is assumed to be 44 (i.e. total species in the three samples). The 2 × 2 tables for the three sample pairs are as follows:

1. Creek zone vs. wash zone

	Creek zone	
Wash zone	Species present	Species absent
Species present	a = 11	b = 12
Species absent	c = 12	d = 9

2. Creek zone vs scarp zone

	Creek zone	
Scarp zone	Species present	Species absent
Species present	a = 15	b = 9
Species absent	c = 13	d = 7

3. Wash zone vs. scarp zone

	Wash zone	
Scarp zone	Species present	Species absent
Species present	a = 10	b = 14
Species absent	c = 13	d = 7

The coefficients are calculated using formulae given in the text as follows: Coefficient of Jaccard, S_j (14); Coefficient of Sorensen, S_s (15); Simple matching coefficient, S_{sm} (16); Baroni-Urbani and Buser, S_B (17). The calculated coefficients are shown below:

Coefficient	Sample comparison		
	Creek vs Wash zone	Creek vs Scarp zone	Wash vs Scarp zone
S_j	0.314	0.484	0.270
S_s	0.478	0.652	0.426
S_{sm}	0.432	0.628	0.386
S_B	0.466	0.640	0.343

Note that all the coefficients show greater similarity between the creek and scarp zone miombo than any of these two with the wash zone miombo and that the coefficient of Jaccard consistently gave the lowest similarity values.

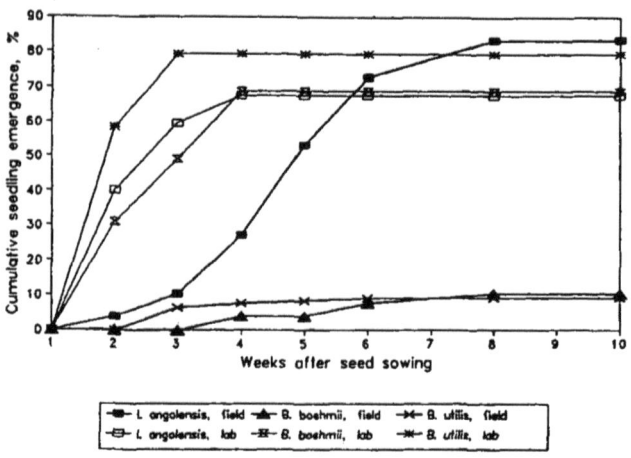

Figure 4.11 Pattern of seed dispersal by an isolated Julbernardia globiflora tree

Table 4.10 Annual variations in seed production in two tree species at four central Zambian dry miombo sites

Year	Woodland type	Site	Estimated seed production (number ha^{-1})	
			Isoberlinia angolensis	Julbernardia globiflora
1990	Regrowth	9-yr old	260	1 270
		18-yr old	0	1 810
	Old growth	A	900	3 330
		B	340	15 400
1991	Regrowth	10-yr old	1 650	0
		19-yr old	0	0
	Old-growth	A	4 720	900
		B	1 820	4 130
1992	Regrowth	11-yr old	0	0
		20-yr old	0	0
	Old-growth	A	1 240	4 757 750
		B	480	4 970 500
1993	Regrowth	12-yr old	2 260	0
		21-yr old	0	0
	Old-growth	A	12 520	970
		B	4 810	4470

Based on Chidumayo (1993a)

There is considerable annual and spatial variations in seed production, especially in old-growth miombo. In most years, production at sites studied by Chidumayo (1993a) was < 0.5 seed m^{-2}. This low seed density implies that seed censuses should be done in a large number of small (< 1.0 m^2) quadrats, or fewer but larger (> 1.0 m^2) quadrats to obtain reliable estimates. A census of *J. globiflora* seeds in 27 small (1.0 m^2) non-random

Table 4.11 *Seedling emergence rate from seeds of miombo trees sown in miombo woodland soil immediately after seed ripening or dispersal*

Species	Seedling emergence rate (%)	
	Field conditions	Laboratory conditions
Afzelia quanzensis	99	86
Albizia antunesiana	40	33
Brachystegia boehmii	10	69
Brachystegia spiciformis	90	83
Brachystegia utilis	30	79
Isoberlinia angolensis	74	73
Julbernardia globiflora	73	73
Julbernardia paniculata	67	69
Pterocarpus angolensis	25	–
Swartzia madagascariensis	36	–
Uapaca kirkiana	43	95

Based on Chidumayo (1993a)

quadrates at an old-growth dry miombo site gave a mean of 4.11 m^{-2} compared to that of 497 estimated from tree density, fruiting frequency, fruits per tree and seeds per fruit data. The disparity is caused by irregularity in tree distribution and seed dispersal (Figure 4.11): the highest seed concentration occurs near tree bases. Because of this, seed dispersal in miombo is rarely uniform in space. Data obtained by the Zambia National Council for Scientific Research (unpublished) indicate that *U. kirkiana* can produce 94 575 seeds ha^{-2}.

Field and laboratory studies are necessary to understand various aspects of seed germination and seedling development and mortality. The source or provenance of seeds for such studies should be recorded together with the date of collection. Seeds should be sown on the soil surface or by burying under a thin soil layer (< 5 cm), preferably in rows at a specified interval for the easy identification of seedlings.

Seed germination rates, as determined by seedling emergence among miombo trees, range from 10–90 per cent (Table 4.11). Some species germinate differently under field and laboratory conditions (Figure 4.12). Chidumayo (1993a) found that dried seeds of *I. angolensis* failed to germinate under field conditions and only 20 per cent germinated in the laboratory. Apparently, this species minimizes the danger of seed desiccation under natural conditions by dispersing some seeds at the beginning of the rainy season. The seed germination period in the majority of miombo trees ranges from 2–6 weeks, although this may be extended to 10 weeks (Figure 4.12), and the germination rate under field conditions may be lower than under laboratory conditions due to mortality during the germination period induced by dry spells at the beginning of the rainy season (Strang 1966). Only seeds of *Pterocarpus angolensis* are known to germinate during the second and even third rainy season, after surviving through the previous rainy seasons following dispersal (Boaler 1966). The majority of miombo trees seeds die and decompose if they fail to germinate during the first rainy season, and this is the reason for the scarcity of seeds in miombo at the end of the rainy season.

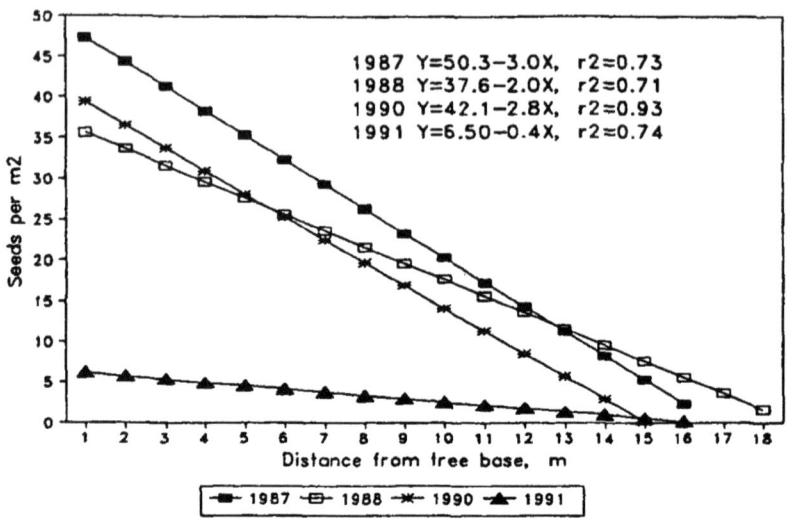

Figure 4.12 *Seedling emergence in three miombo trees*

Seedlings from seeds that germinate during the same season form a cohort which is the basic unit for studying survivorship (Figure 4.13). Survivorship is the number of individuals in a cohort that are alive at a given age, expressed as a percentage. Seedling development can be studied by harvesting seedlings of a known age and determining the shoot and root (at a given depth) biomass. Non-destructive methods include measuring shoot height and estimating leaf area from the number and size of leaves.

Table 4.12 *Trend in deforestation caused by cultivation in Zambia during 1969–1990 and projections for the years 2000 and 2010*

		Semi-permanent and permanent cultivation		Chitemene shifting cultivation		Total deforestation (1000 ha)
Year	Total h/holds (* 1000)	Land under cultivation (1000 ha)	Annual deforestation (1000 ha)	Total h/holds (*1000)	Annual deforestation (1000 ha)	
1969	480	1 093	na	104	432	na
1950	546	1 244	152	119	492	644
1990	658	1 501	256	143	593	849
2000	760	1 700	230	160	680	910
2010	950	2 200	450	210	860	1 300

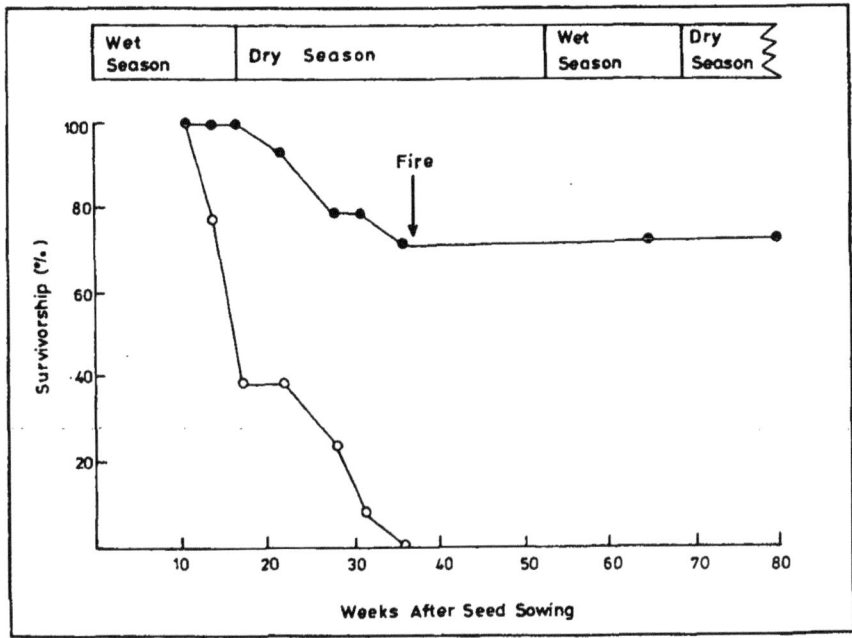

Figure 4.13 *Initial survivorship of seedlings of Isoberlinia angolensis* (dots) and Uapaca kirkiana (bullets) *seedlings at a dry miombo site*

Cultivation and deforestation

In Zambia there are basically two systems of cultivation: semi-permanent/ permanent and shifting (chitemene). Assessment of deforestation by cultivation has been based on the assumption that each rural household has a cultivation plot and that an increase in households results in a corresponding increase in cultivation plots through conversion from forest land. Annual deforestation caused by semi-permanent or permanent cultivation can therefore be estimated by the product of the increase in rural households and the average cultivated farm plot. Deforestation resulting from chitemene cultivation can be estimated by the product of the total households in the chitemene farming system in a given year and the annual forest requirement. Data on households can be obtained from national censuses.

In estimating deforestation caused by semi-permanent/permanent cultivation system in Table 4.12, the average farm plot size is assumed to have remained fairly constant at 2.28 ha per household (Schultz 1974) during the 1964–94 period. This is a conservative assumption, because there is evidence that average farm plot size increased by about 2.8 per cent per annum from 2.15 ha in 1970 (Schultz 1974) to 3.40 ha in 1989 (Kwesiga and Chisumpa 1992). In Central and Southern Provinces the increase in average farm plot size may have been even higher: from 3.00 ha in 1970 (Schultz 1974) to 4.14 ha in 1976 (Marter and Honeybone 1976). Among

Lima women farmers in Choma and Mumbwa districts the average farm size was between 9 and 19 ha in 1985 (Chilivumbo and Kanyangwa 1985). In addition, usually the area under cultivation is less than the actual farm-holding, since semi-permanent or permanent cultivation implies resting part of the farm holding each year. In Eastern Province, Kwesiga and Chisumpa (1992) found that the area under cultivation represented 43 per cent of the average farm holding of 6.30 ha in 1989. Thus the use of the area under cultivation may underestimate deforestation caused by semi-permanent or permanent cultivation.

The average woodland clearing for chitemene shifting cultivation has remained constant at 8.3 ha per household during 1964–94 (Stromgaard 1989). However, the frequency of new clearings for chitemene has decreased from yearly to once in two years (see Table 2.3) and the estimate of annual deforestation caused by chitemene in Table 4.12 was therefore based on half the average clearing (8.3/2 = 4.15 ha) per household.

Critical population density under shifting cultivation

A suitable measure of the carrying capacity under shifting cultivation was developed by Allan (1949):

$$\text{CPD} = \left(\frac{100}{C_p}\right) C_a \times L \qquad \text{Critical population density (19)}$$

where CPD is the critical population density, C_p is the percentage of land which is cultivable, C_a is the area required to allow a proper ratio of the years of fallow to the number of successive years of cultivation. L in (19) is calculated according to 20.

$$L = \left(\frac{R}{U}\right) + 1 \qquad \text{Fallow ratio (20)}$$

where R is the number of years of fallow and U is the number of successive years of use of a cultivated plot.

Table 4.13 *Land categories and use in the chitemene cultivation region in Zambia*

Land category/use	Area (km²)
Uncultivable land:	
Forest reserves	6 959
National parks	3 898
Hilly area	17 300
Wetlands	15 891
Cultivable land:	
Cropland and fallow	59 283
Unused woodland	38 756
Total:	142 087

Based on Schultz (1974)

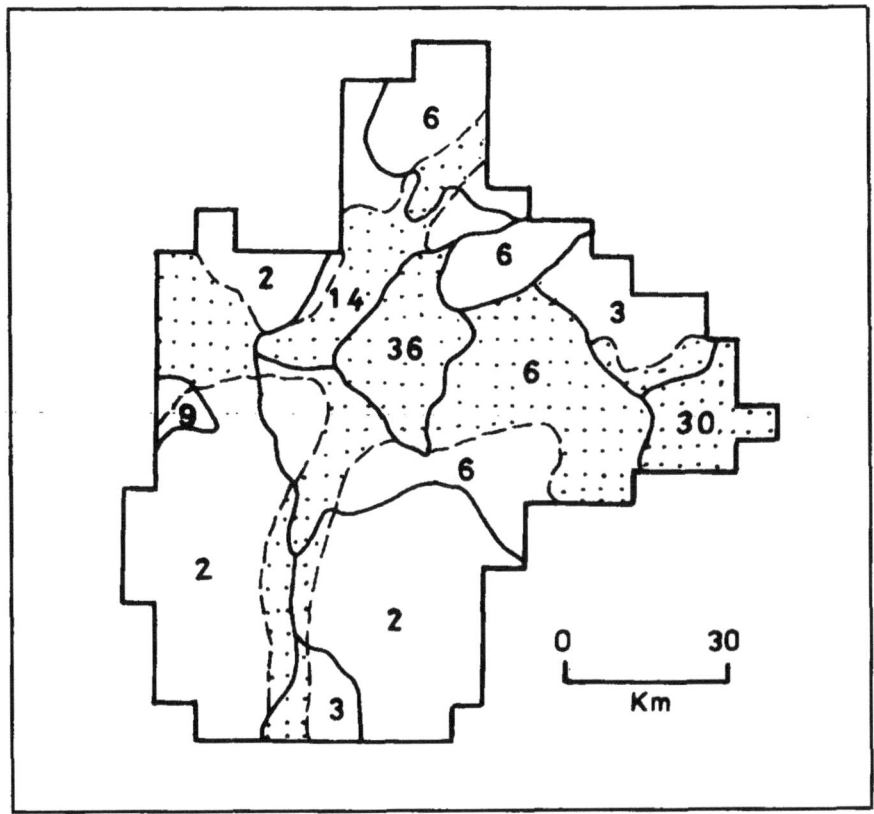

Figure 4.14 *Correlation between population density (numbers) and deforestation caused by chitemene shifting cultivation (shaded area) in Mansa district, Luapula Province, Zambia*

The uncultivable land includes national parks, forest reserves, hilly areas and wetlands which should be subtracted from the total land area to derive the cultivable land (Table 4.13). The percentage cultivable land is then calculated as follows:

$$C_P = \left(\frac{\text{Total area} - \text{uncultivable land}}{\text{Total area}} \right) \times 100 \quad \text{Percentage cultivable land (21)}$$

In the case of the chitemene shifting cultivation region in Zambia (Table 4.10) the C_p is 69 per cent.

In 1990 the average rural household size was 5.3 (based Table 2.1) and given an annual woodland clearing of 4.15 ha, the area required to support one person per year (C_a) was 0.78 ha. The average AG wood biomass of 30 t ha^{-1} harvested for chitemene (Araki 1992) can regenerate in 15 years (see Figure 1.14) which is the 'R' in (20). The number of successive years of use

> **Box 4.7** *Problems associated with the critical population density concept for chitemene shifting cultivation in Zambia.*
>
> Chitemene shifting cultivation has been described in Chapter 2 and the issue of carrying capacity relates to the question of whether chitemene can survive under population pressure. Chitemene systems in Zambia are dynamic and evolve in response to a variety of factors, both socioeconomic and ecological (Trapnell 1953).
>
> The estimation of critical population density (CPD) as a measure of carrying capacity for chitemene is currently beset with a number of problems. Firstly chitemene cultivators do not rely entirely on their chitemene fields for all their requirements. The cultivators also cultivate village gardens and employ other subsistence strategies, such as fishing, hunting and gathering to obtain some of their requirements. The calculation of CPD does not take into account these other subsistence strategies which permit the maintenance of higher population densities than could be maintained by a chitemene system alone.
>
> Secondly, there are local variations in the percentage of land that is suitable for chitemene production (C_p). Estimates range from 50–70 per cent. Calculation of CPD is based on the assumption that the population is distributed evenly throughout the cultivable land. This is usually not the case, as both political policies and infrastructure development (i.e. roads and community services) affect population distribution.
>
> Thirdly, there is inadequate knowledge about woodland regeneration which is the basis for determining the suitable fallow period (R). Both the size of trees and total wood biomass are important factors in determining R and the regeneration of woodland is also affected by fire management and site quality which may vary.
>
> The estimation of carrying capacity for chitemene therefore only provides a very rough guide about the man–land relationship and can vary considerably depending on the magnitude of the input variables. This is clearly shown below by using different values for C_p and R in calculating CPD for chitemene.
>
> | Percent cultivable land (Cp) | Critical population density (persons km^{-2}) | | | | |
> |---|---|---|---|---|---|
> | | Woodland regeneration period (R) in years | | | | |
> | | 30 | 25 | 20 | 15 | 10 |
> | 50 | 2.07 | 2.47 | 3.05 | 4.01 | 5.83 |
> | 55 | 2.27 | 2.71 | 3.36 | 4.41 | 6.41 |
> | 60 | 2.48 | 2.96 | 3.66 | 4.81 | 6.99 |
> | 65 | 2.69 | 3.21 | 3.97 | 5.21 | 7.58 |
> | 70 | 2.89 | 3.45 | 4.27 | 5.61 | 8.16 |

of a cultivated plot for finger millet, the 'U' in (20), is 1. The number of plots required to allow a proper ratio of the years of fallow to the number of successive years of use in 1990 can be calculated using (20)

$$L = \left(\frac{15}{1}\right) + 1 = 16$$

The critical population density in 1990 in the chitemene land use region can then be calculated using (19):

$$CPD = \left[\left(\frac{100}{69}\right) \times 0.783\right] \times 16 = 18 \text{ ha}$$

This CPD represents a population density of 5.56 km^{-2}. It follows therefore that in areas where this CPD has been exceeded, severe deforestation can be expected (Figure 4.14). Moore and Vaughan (1994) have questioned the usefulness of the CPD for chitemene on the basis that the CPD concept is dynamic and its calculation is founded on simple assumptions, while factors that determine the CPD are complex. These issues are discussed in Box 4.7.

Charcoal yield and deforestation
Charcoal yield in earth kilns is calculated as follows:

$$\text{Charcoal yield} = \frac{\text{Charcoal yield}}{\text{Wood carbonised}} \qquad \text{Charcoal yield (24)}$$

The charcoal produced consists of bagged and unbagged (residual) charcoal (see chapter 2). Bagged charcoal is estimated by multiplying the number of bags by the average mass of charcoal per bag (41 kg for a 90-kg grain bag). On average, bagged charcoal is 97 per cent of the charcoal produced (Chidumayo 1994a). Charcoal produced is therefore estimated by dividing the weight of bagged charcoal by 0.97. On average carbonized wood makes up 96 per cent of the total wood in the kiln and mass of carbonized wood is estimated by multiplying total wood in the kiln by 0.96. Average charcoal yield rate in earth kilns on a oven-dry weight basis is 23.4 per cent ($SD = 7.2$). This conversion rate gives a yield of 0.23 kg charcoal per kg of carbonized wood, while the output of bagged charcoal is 0.22 kg per kg of carbonized wood.

A number of household energy demand surveys have been done in Zambia since 1978. Urban surveys were undertaken in 1978 (Chidumayo 1979), 1983 (Chidumayo and Chidumayo 1984), 1988 (World Bank 1990) and 1994 (Macwani *et al.* 1994). Each of these surveys were carried out in a number of urban areas. Fewer rural surveys have been undertaken in Chipata, Lusaka Rural and Mongu (Chidumayo 1985 and unpublished) and Zambezi (Christensen 1985). The FAO/Forest Department survey of 1986 obtained data on household biomass energy consumption in both urban and rural areas, and the World Bank/Department of Energy survey of 1988/89 was the only comprehensive survey that collected data on biomass energy consumption in all sectors, including commerce and industry. The estimation of wood used in charcoal production is based on a calculation from the amount of charcoal consumed and the charcoal yield rate in earth kilns.

The basic charcoal consumption unit in the domestic sector is the household, while per capita consumption estimates are derived from household consumption data. Cline-Cole *et al.* (1990) showed that household size is an important determinant of fuelwood demand at the household level. In Zambia there is a positive correlation between household size and charcoal consumption (r-squared = 0.86) while the correlation between per capita consumption and household size is negative (r-squared = 0.78). This means that although biomass energy consumption increases as household size increases, per capita consumption declines (Figure 4.15). It is therefore important to understand changes in average household size and the number of households (Table 4.14) before estimating charcoal energy consumption over time.

Table 4.14 *Changes in households and average household size in Zambia*

Year	Total households			Mean household size		
	Urban	Rural	Total	Urban	Rural	Total
1969	243 265	622 826	866 091	4.9	4.6	4.7
1980	382 712	709 167	1 091 879	5.9	4.8	5.2
1990	547 667	855 283	1 402 950	6.0	5.3	5.6
2000	738 033	983 390	1 721 423	6.1	5.9	6.0
2010	994 839	12 37 833	2 232 672	6.2	6.0	6.1

Based on Chidumayo (1994a)

Mean household size in Zambia increased during 1969–90, and will probably continue to do so up to 2010 and beyond. By implication (see Figure 4.15), average charcoal consumption per household also increased, while per capita consumption decreased. Fortunately, these changes were not large. For example, gross annual charcoal consumption per household averaged at 1043 kg in 1983 (Chidumayo and Chidumayo 1984), 984 kg in 1988 (World Bank 1990) and 1110 kg in 1994 (based on the 1994 survey data). Comparison between the years 1988 and 1994 did not reveal significant differences ($t < 1.50$; $P > 0.05$). The average annual charcoal consumption in urban Zambia during 1983–94 was therefore 1046 kg per household. Incidentally, the proportion of urban households using charcoal remained at about 85 per cent during 1978–94.

Since average charcoal consumption in urban areas remained fairly constant, total household charcoal consumption can be estimated by multiplying the number of households by average annual consumption per household. From the available data, average charcoal consumption per household in rural Zambia is estimated at 100 kg per year. The household sector accounts for 96 per cent of total charcoal consumption in the country (Department of Energy 1992). Total charcoal consumption can therefore be estimated by dividing charcoal consumption in the household sector by

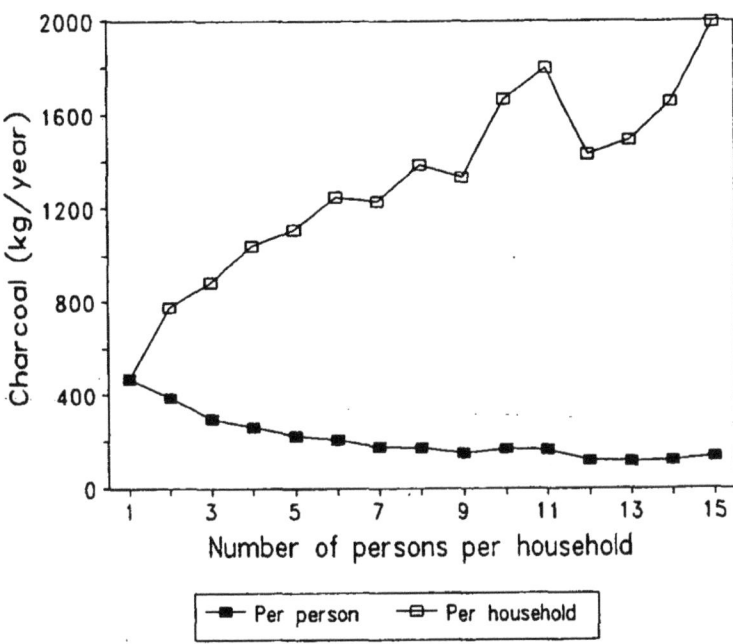

Figure 4.15 *Correlation between household size and annual charcoal consumption per household and per capita in urban Zambia*

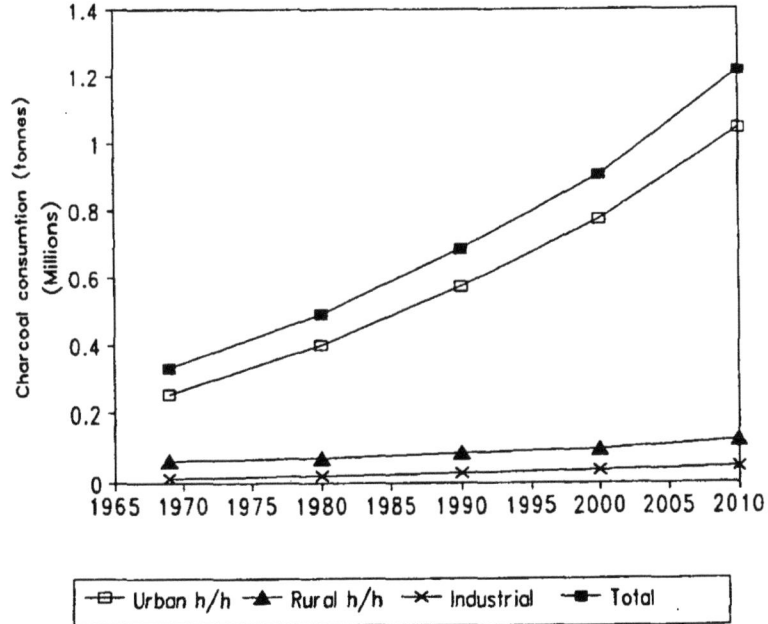

Figure 4.16 *Increase in charcoal consumption in Zambia*

0.96. Figure 4.16 shows trends in charcoal consumption based on the growth in households, and a constant average consumption per household calculated separately for rural and urban areas. On the basis of these trends, charcoal consumption increased from 0.32 million tonnes in 1969 to 0.47 million in 1980 and 0.66 million in 1990, and is projected to increase to 0.87 million and 1.16 million tonnes in 2000 and 2010, respectively. The wood used in charcoal production in Zambia can be calculated as in (25).

$$\text{Wood used} = \frac{\text{Charcoal produced}}{\text{Charcoal yield}} \qquad \text{Wood use for charcoal production} \quad (25)$$

where charcoal produced equals charcoal consumed/0.97, and yield rate is 0.23.

Table 4.15 shows wood used in charcoal production in 1969–90 and projections for 2000 and 2010.

Table 4.15 *Charcoal consumption and wood used in charcoal production in Zambia*

Year	Charcoal consumption (million tonnes)	Charcoal produced (million tonnes)	Wood used (million tonnes)
1969	0.330	0.340	1.179
1980	0.490	0.505	2.196
1990	0.685	0.760	3.070
2000	0.905	0.933	4.056
2010	1.211	1.248	5.428

From Chidumayo (1994a)

Once the total wood used in charcoal production is estimated, deforestation can be calculated from the average cord wood biomass per hectare, and the proportion used for charcoal production by the earth-kiln method (Table 2.11). In miombo the proportion of cord wood used for charcoal production is 97 per cent. If this proportion is applied to other vegetation types in conjunction with data Table 1.13, the estimated charcoal production per hectare would be as given in Table 4.16. By dividing either total charcoal produced by charcoal produced per hectare or total wood used for charcoal production by harvested cord word per hectare, deforestation through charcoal production can be estimated. For example, if all the wood used for charcoal production in 1990 (Table 4.15) came from dry miombo, deforestation would be estimated at 54 820 ha while the estimate for wet miombo would be 41 490 ha.

Clearly this method of estimating deforestation caused by charcoal production involves a number of problems. Firstly the method assumes that all the wood used in charcoal production is the primary cause of forest clearing. This is not always the case, because charcoal is also produced from

Table 4.16 *Estimated charcoal production per ha by the earth kiln method in different vegetation types in Zambia*

Vegetation type	Above ground cord wood		Charcoal produced (t ha^{-1})
	Total	Harvested	
Dry evergreen forest	158	153	36
Miombo: wet	76	74	17
dry	58	56	13
kalahari	43	42	10
Mopane/Munga	38	37	9

wood in areas cleared primarily for agriculture. The Kanakantapa settlement area north-east of Lusaka was a significant source of charcoal for Lusaka, the charcoal being produced from land clearing for agriculture (Kalumiana 1994). Unfortunately the amount of charcoal produced from agricultural land clearing in Zambia is not known.

Secondly, our knowledge of how much wood is cut and the proportion actually used for charcoal production is very scanty. This affects the reliability of estimates of deforestation. In the Chisamba area, north of Lusaka, Chidumayo (1991b) found that 94 per cent of the AG cord wood biomass was cut and only 92 per cent was actually carbonized (see Table 2.11). The use of total cord wood or cord wood cut per hectare can therefore underestimate deforestation.

Thirdly, the great spatial variation in forest biomass (see Table 4.13), even at local scale, affects the accuracy of estimates of deforestation caused

Table 4.17 *Assessments of early and late burning on stem mortality after two years of treatment at Mutupa plots in the Zambian Copperbelt*

	Coppiced plots		Woodland plots	
Variable	Early burnt	Late burnt	Early burnt	Late burnt
Albizia adianthifolia				
Total stems	212	221	28	18
Dead stems	25	101	2	11
Mortality rate (%)	11.8	45.7	7.1	61.1
Baphia bequaertii				
Total stems	168	114	27	21
Dead stems	1	47	0	6
Mortality rate (%)	0.6	41.2	0.0	28.6
Marquensia macroura				
Total stems	109	174	39	29
Dead stems	12	108	1	4
Mortality rate (%)	11.0	62.1	2.6	13.8
Uapaca kirkiana				
Total stems	77	38	37	35
Dead stems	1	13	1	6
Mortality rate (%)	1.3	34.2	2.7	17.1

Based on Chidumayo (1989a)

> **Box 4.8** *Problems of determining the spatial pattern of deforestation in Zambia.*
>
> The approach used to determine deforestation in Zambia has largely been based on a backwards calculation from annual demand for land for cultivation and charcoal production. This approach cannot be used to determine the spatial pattern of deforestation, and yet knowing where deforestation is occurring is important for managing the problem. Both remote-sensed data (i.e, satellite imagery and aerial photographs) and ground-truthing are necessary for analysing the spatial pattern of deforestation. Individual patches of annual forest clearing are usually small, but they expand and coalesce over a period of years when the problem becomes apparent. Time-series data are therefore necessary in assessing spatial trends in deforestation. The time interval between data sets should not be too long for regeneration to obscure the actual trend. When this occurs the problem can only be resolved by recourse to historical data on land use.
>
> This type of assessment was attempted by Chidumayo (1989b) and Chidumayo and Chidumayo (1984), who used a combination of aerial photographs, archival information and field visits to determine both temporal and spatial patterns of deforestation in the Copperbelt area of Zambia as shown in Figure 4.17. The acquisition of time-series aerial photographs and ground-truthing can be expensive. Satellite imagery is very useful in monitoring deforestation and land use changes and can be very cost-effective, but it often requires special skills and equipment for correct interpretation. For these reasons little has been done to determine spatial trends in deforestation in Zambia, and consequently this has hampered the development of realistic strategies for managing the problem of deforestation.

by charcoal production. With all these uncertainties, estimates of deforestation should be used carefully. Fourthly, determining areas where deforestation occurs has remained problematic (Box 4.8).

The effects of fire

Experiments to investigate the effects of wild fires have been carried out in many parts of Africa (Brookman-Amissah *et al.* 1980). Sets of sample plots are demarcated and enumerated and then subjected to different burning regimes: early burning, late burning and complete protection. Repeated assessments in subsequent years allow the effects of fire to be quantified. Where the experiments include coppicing as a silvicultural technique, trees should be measured and recorded before the experiment and stumps marked in order to monitor stump mortality. For the assessment of tree growth, mortality and recruitment, trees must be tagged or marked, and periodic girth or diameter measurements made. Many previous experiments have used tree girth or height classes and these classifications have precluded the accurate assessment of tree growth, mortality and recruitment. However, few experiments have been adequately maintained and the effects monitored over sufficient number of years.

> **Box 4.9** *Using change in relative importance to assess response of miombo woodland trees to fire management.*
>
> Trapnell (1959) and Lawton (1978) used numerical abundance under different fire treatments to determine fire-tolerance of miombo woodland trees (see Table 3.7). This approach can give misleading results if post-treatment abundance is not compared to pre-treatment abundance. The use of changes in relative importance of species before and after fire treatment overcomes this problem. The rationale of such an analysis is that if all the species are affected in the same way by fire, there should be no change in their relative importance after fire treatment. The significance of the change in relative importance can also be tested statistically using the difference in proportions (Hayslett 1967) which further enhances the interpretation of results. The formula in (26) is used in the analysis.
>
> At a confidence level of 0.05 the z-value is 1.645. If the calculated value of z is greater than 1.645 or less than −1.645 then the species changed significantly due to fire treatment. Negative z values indicate beneficial response (an increase in importance) while positive values indicate fire intolerance (a decrease in importance). Species that are fire neutral (no significant change in importance) have z values greater than −1.645 and smaller than 1.645.
>
> This analysis was carried out using data in Appendices 9 and 10. Eleven years of fire treatments had no significant effect on all the species in old-growth woodland. However, a combination of clear cutting and fire treatments yielded different results, as shown in Table 4.18.
>
> The majority of species showed no significant change in relative importance under early burning, except *E. africanum, I. angolensis* and *P. curatelifolia* which responded negatively (decreased in importance) during the early regeneration phase. *S. innocua* and *S. pungens* responded negatively during the late regeneration phase. Species such as *A. boehmii, B. bequaertii, B. longifolia* (late regeneration phase) and *D. condylocarpon* (early regeneration phase) responded negatively to fire protection. Only *J. paniculata* responded negatively to both early burning and protection throughout the regeneration phases. Many of the small trees and shrubs, such as *Bridelia duvigneaudii, Byrsocarpus orientalis, Canthium crassum, Chrysophyllum bangweolense, Hexalobus monopetalus, Maprounea africana, Memycylon flavovirens, Ochna schweinfurthiana, Psorospermum febrifugum, Randia kuhniana, Strychnos cocculoides, S. spinosa* and *Xylopia adoratissima* increased in importance during the early regeneration phase but disappeared during the late regeneration phase. It is not clear whether their disappearance is a consequence of shading by canopy species, or their short life span.

Given the great spatial diversity of miombo (see Chapter 1.2), it is not easy to replicate plots unless these are small (< 0.1 ha). The layout of experimental treatments should therefore take into account local environmental gradients. To confine sample plots to a particular treatment, these should be separated by sufficiently wide (2–5 m) perimeter firebreaks. Assessments of responses to different fire treatments can be based on comparisons of proportions of total trees or relative changes in abundances before and after the experiment (see Box 4.9) and/or between treatments (Table 4.17). The formula used in the analysis is presented as equation (26)

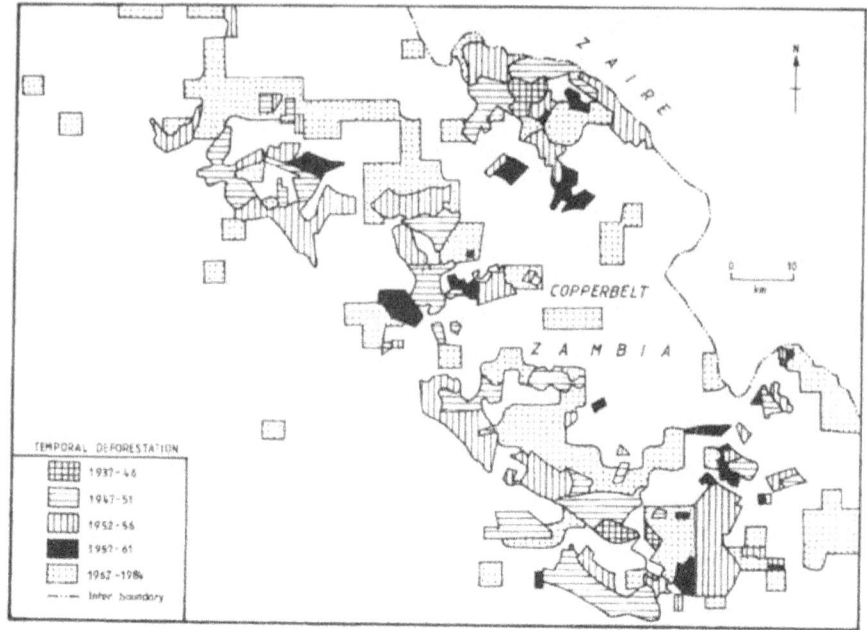

Figure 4.17 *Spatial and temporal patterns of deforestation in the Copperbelt*

$$z = \frac{(p_1 - p_2)}{\sqrt{\left[\frac{p_1(1-p_1)}{n_1}\right] + \left[\frac{p_2(1-p_2)}{n_2}\right]}} \qquad \text{Statistic for sample proportions (26)}$$

where p_1 = proportion of species i before treatment
        p_2 = proportion of species i after treatment
        n_1 = total individuals of all species before treatment
        n_2 = total individuals of all species after treatment
        $p_1 n_1 > 5$, $p_2 n_2 > 5$

Comparisons of mortality rates in Table 4.17 indicate that (i) late burning is more damaging to trees than early burning, (ii) *A. adianthifolia* died more frequently under either type of burning than the other three species in the woodland and (ii) *A. adianthifolia* and *M. macroura* in the coppiced plots suffered more damage under early burning than *B. bequaertii* and *U. kirkiana*. Thus *M. macroura* requires fire protection in coppiced plots, while *A. adianthifolia* requires protection in both woodland and coppiced plots. For *B. bequaertii* and *U. kirkiana*, early burning is sufficient as a silvicultural technique for their management.

Through differential impacts on miombo woodland trees, fire

management has implications for species diversity conservation. Species diversity analysis using data in Appendices 9 and 10 revealed that after a decade of fire management in old-growth wet miombo, species diversity decreased under all the three fire management regimes (Table 4.17). In regrowth areas, species diversity increased during the early regeneration phase but decreased during the late regeneration phase. However, late burning maintains the regrowth in an early regeneration phase in which species diversity remains fairly high.

Most fire experiments have considered either trees or herbs. Changes in herbage production should be related to pre-treatment levels. This may require assessments for several years before initiating the experiments.

Table 4.18 *The response of miombo species to clear cutting and fire management*

Species	After 11 years		After 49 years	
	Protected	Early burnt	Protected	Early burnt
Albizia antunesiana and A. Adianthifolia	None	None	Increase	Increase
Anisophyllea boehmii	Decrease	None	Decrease	None
Baphia bequaertii	Decrease	None	Decrease	None
Brachystegia longifolia	None	–	Decrease	–
Brachystegia spiciformis	Increase	None	Increase	None
Combretum molle and C. zehyeri	None	None	None	None
Dialiopsis africana	Increase	None	–	None
Diospyros botacana	None	None	–	None
Diplorhynchus condylocarpon	Decrease	None	None	None
Erythrophleum africanum	None	Decreased	Increase	None
Isoberlinia angolensis	Decrease	Decreased	Increase	Increase
Julbernardia paniculata	Decrease	Decreased	Decrease	Decrease
Lannea discolor	None	None	None	None
Pericopsis angolensis	None	–	None	–
Phyllocosmus lemaireanus	Increase	None	–	None
Pseudolachnostylis maprouneifolia	Increase	Increase	Increase	Increase
Strychnos innocua and S. pungens	Increase	None	None	Decrease

Table 4.19 *Changes in species diversity in wet miombo woodland under fire management at Ndola, Zambia*

Woodland		Shannon-Wiener species diversity index		
		Protected plot	Early burnt plot	Late burnt plot
Old growth:	1933	1.341	1.353	1.373
	1944	1.244	1.198	1.242
Regrowth: before cutting in	1933	1.025	0.988	0.983
	1944	1.428	1.426	1.268
	1982	1.075	1.142	–

Based on data in Appendices 9 and 10

Where this is not possible herbage production can be compared between adjacent control and experimental plots. Successional and soil changes can be similarly assessed. Wild fires in dry miombo appear to cause a decrease in both carbon and total nitrogen at the soil surface and an increase in carbon in the subsurface soil immediately after their occurrence (Table 4.20).

Table 4.20 *The immediate effect of an August wild fire on a dry miombo soil at Chisamba in central Zambia*

Parameter	Mean element content (% of dry soil)			
	0–2 cm soil depth		2–10 cm soil depth	
	Before fire	After fire	Before fire	After fire
Carbon	1.74	1.51	0.85	0.95
Nitrogen	0.11	0.09	0.07	0.07

Based on Ward (unpublished)

Very few experiments have assessed the efficiency of fires. This can be done by measuring the pre-fire and post-fire biomass fuel in adjacent micro-plots along a transect. A large sample (> 30) of micro-plots is necessary to minimize the effects of spatial heterogeneity (Shea *et al.* 1993). Fire efficiency is then calculated as follows:

$$\text{Fire efficiency (\%)} = 100 - \left[\frac{FB_{\text{post-fire}}}{FB_{\text{pre-fire}}}\right] \times 100 \qquad \text{Fire efficiency (27)}$$

where FB is mean fuel biomass m^{-2} before (pre-fire) and after (post-fire) fire.

The effects of wood carbonization on soil
The effects of wood carbonization on the soil at earth kilns (see chapter 2) can ideally be assessed by collecting samples at given depths before and after carbonization. Where this is not possible, soil samples from the kiln site can be compared with those from the adjacent undisturbed area. The samples are analysed for key soil characteristics. If the objective is to monitor the persistence of the effects, repeated samples should be periodically collected at the same soil depths and sites. The results can be compared between the kiln and undisturbed sites and on each site over time. This method was used by Chidumayo (1993a, 1994c), who found that wood carbonization increased soil pH and exchangeable P (Table 4.21) and K.

The effects of deforestation
Experiments to assess the impact of deforestation on catchment hydrology involve the measurement of vegetation cover, stream flow, soil moisture,

Table 4.21 The changes in the concentration of extractable phosphorus in charcoal kiln soil in dry miombo in central Zambia

	Mean extractable phosphorus content (mg kg^{-1} soil)	
Soil depth (cm)	Undisturbed soil	Charcoal kiln soil
0–10 (SD)	10.6 (11.2)	29.4 (9.0)
11–30 (SD)	3.4 (1.7)	27.7 (7.2)

Based on Chidumayo (1994c)

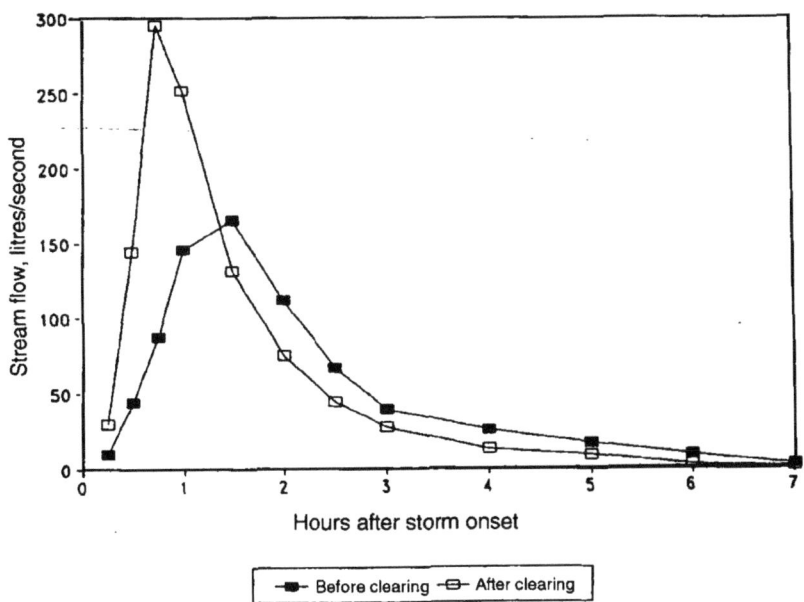

Figure 4.18 Mean stream flow and duration before (1969–73) and after clearing wet miombo (1975–79). Based on Mumeka 1986.

ground water, rainfall, evaporation, wind speed, air temperature, sunshine hours and relative humidity. Data on the latter six variables can be obtained from weather stations in or near the experimental catchment. Since no two catchments are likely to be the same, experiments to assess the impact of deforestation should have two phases. The first phase should be designed to establish baseline characteristics and the pre-disturbance behaviour of the catchments. The second phase should monitor the effects of deforestation treatments. For details on such experiments, the reader is referred to Pereira (1975), Amphlett (1986) and Hudson (1984).

The only available such study in miombo was conducted by the National Council for Scientific Research in wet miombo. The study assessed the combined effects of deforestation (removal of 80–85 per cent of woodland cover) and subsistence agriculture in the Zambian Copperbelt (Mumeka 1986). The effects of these treatments on stream flow are shown in Figure 4.18. Deforestation in miombo increases yearly runoff by 10–18 per cent, while the decrease in evapotranspiration due to canopy removal increases base flow and ground water storage (Sharma 1985). The immediate effect of woodland clearing is an increase in peak stream flow and the shortening of flow duration (Figure 4.18). Clearing miombo also increases grass production (Chidumayo 1993a; Hood 1972) but this has little effect on catchment hydrology because herbaceous plants contribute little to evapotranspiration during the dry season (see Chapter 1).

Table 4.22 *Average soil moisture content (% dry weight) in uncut plots and plots clear-cut by stumping in 1990 at an old-growth dry miombo site in central Zambia*

	Mean soil moisture content (% dry weight)			
	October 1992		February 1993	
Soil depth(cm)	Uncut plots	Cut plots	Uncut plots	Cut plots
0–10	1.7	0.9	18.3	16.1
11–30	4.4	2.5	17.7	15.2
31–60	10.2	6.4	18.8	16.9
61–150	7.9	7.1	16.4	17.9

Based on Chidumayo (1993a)

Chidumayo (1993a) carried out a simpler assessment of the effect of clearing trees on soil moisture regime in adjacent control and experimental plots in dry miombo in central Zambia. At these sites coppice regrowth in clear-cut plots used more water in the upper soil layer (0–60 cm) and less in the bottom layer (> 60 cm) than trees in control plots (Table 4.22). Reduced water use in bottom soil layers in cut-over plots would therefore increase ground water storage and base flow, which supports the observations made above.

Assessments of the effects of deforestation on soil characteristics should ideally involve pre-experimental analysis of soil samples to establish baseline conditions, followed by the monitoring of soil conditions after deforestation treatments. Because some soil characteristics show seasonal and annual variations (Chidumayo 1993a), it is important that both pre- and experimental sampling is carried out over a number of years and seasons. No such studies have been done in Zambian miombo, but comparisons of soil characteristics on adjacent control and experimental plots in dry miombo in central Zambia revealed no significant changes in soil nutrient status due to deforestation (Chidumayo 1993a).

Grazing and browsing

Very few studies have assessed grazing and browsing and their effects on miombo. This can be assessed by comparing herbage or browse biomass in sample plots before and after the activity of herbivores. In practice, the assessment of browse biomass in sample plots is more time-consuming and

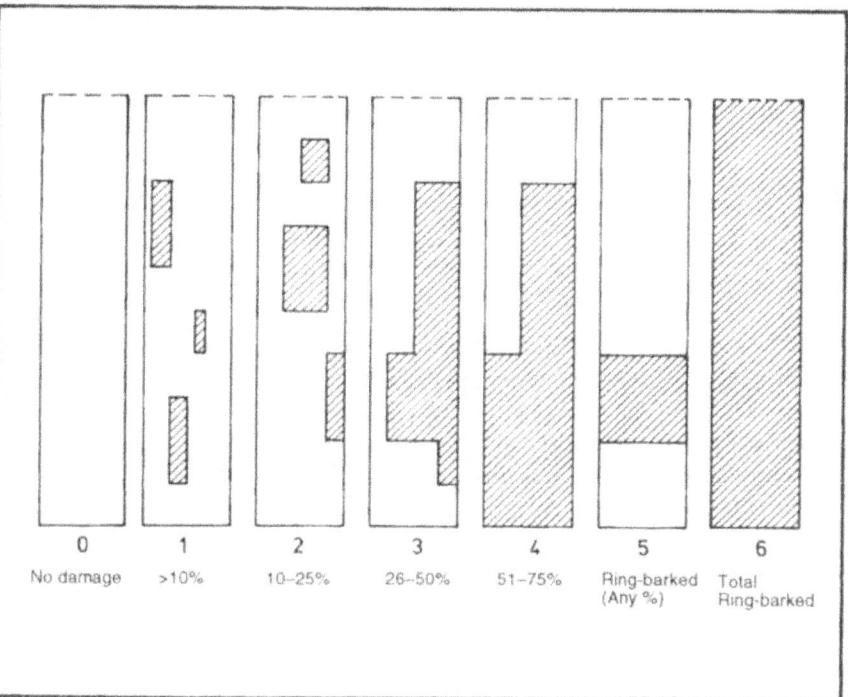

Figure 4.19 *A seven-point scale for assessing bark removal below 2 m (broken line) (Based on Cunningham 1993)*

laborious. It may be preferable therefore to use tagged or marked sample trees, branches and shoots. The effect of larger mammalian herbivores can also be studied by using exclosures. Herbivore pressure is then estimated by calculating the difference in standing plant biomass inside and outside exclosures over the experimental period (Table 4.23). However, in such experiments it is important to assess plant biomass in the exclosure plots before and after the experiment to determine biomass increment during the experiment.

Herbivore pressure is then calculated as follows:

$$HP (\%) = [PB_{ext\ 1} - (PB_{t1} + \{PB_{ext1} - PB_{ext0}\})] \times 100$$

Herbivore pressure (28)

where HP = herbivore pressure
 $PB_{t\ 1}$ = plant biomass after grazing
 $PB_{ext\ 0}$ = plant biomass in exclosure plots before the experiment
 $PB_{ext\ 1}$ = plant biomass in exclosure plots after the experiment

Lewis (1986) assessed the relative rates of elephant browsing by monitoring

Table 4.23 *The effect of mammalian herbivores on a wetland grassland on the Kafue Flats, Zambia*

Sampling date (1980)	Above ground live herbage biomass (g m^{-2})		Herbivore pressure (%)
	Exclosure plots	Unprotected plots	
31 January	284	228	20
21 March	423	278	34
10 December	60	33	45

Based on Ellenbroek (1987)

branch shoot growth and removal on 200 tagged mopane trees and over 573 branches in the Luangwa valley, Zambia. Growth was measured by changes in shoot length and diameter at marked points. He found that browsed branches had a mean decline in shoot length of 14 cm (SD 9.4) during June to October when mopane trees are dormant.

The effects of grazing, such as defoliation and debarking of trees can be graded on a relative scale. Reeler *et al.* (1991) used a six-point rating of defoliation of *Brachystegia spiciformis* by Chrysomelid beetles (*Melasoma quadralineata*): (1) no damage, (2) 1–10 per cent, (3) 11–25 per cent, (4) 26–50 per cent, (5) 51–75 per cent and (6) 76–90 per cent defoliation which were visually assessed. In order to quantify tree damage caused by debarking, Cunningham (1993) proposed a seven-point scale of debarking for field assessments (Figure 4.19).

Litter decomposition

Litter decomposition in miombo has been studied in confined litter bags filled with a known mass of litter which are left in the field (Soil Productivity Research Programme 1987). In situations where litter bags can attract attention and be lost, samples of known quantities of litter are left in microplots either randomly or systematically located on the woodland floor (Chidumayo 1993a; Malaisse *et al.* 1975). The litter mass is estimated by oven-drying a subsample to constant weight at 80 °C to determine MC. The decomposition samples are then retrieved periodically and oven-dried to assess mass loss.

Litter fall may be added to unconfined litter samples in micro-plots during the decomposition period. In order to correct for litter fall input, a separate set of micro-plots should be established at the study site to monitor and assess litter fall (Chidumayo 1993a). Relative litter mass loss is calculated as follows:

$$\text{Litter mass loss (\%)} = 100 - \left\{ \left[\frac{LM_{t\,1} - LFM}{LM_{t\,0}} \right] \times 100 \right\}$$

Litter mass loss (29)

where LM_{t0} and LM_{t1} are litter mass at the beginning and end of the decomposition period, respectively, and LFM is litter mass input from litter fall during the decomposition period. In the case of litter confined in bags LFM is assumed to be zero.

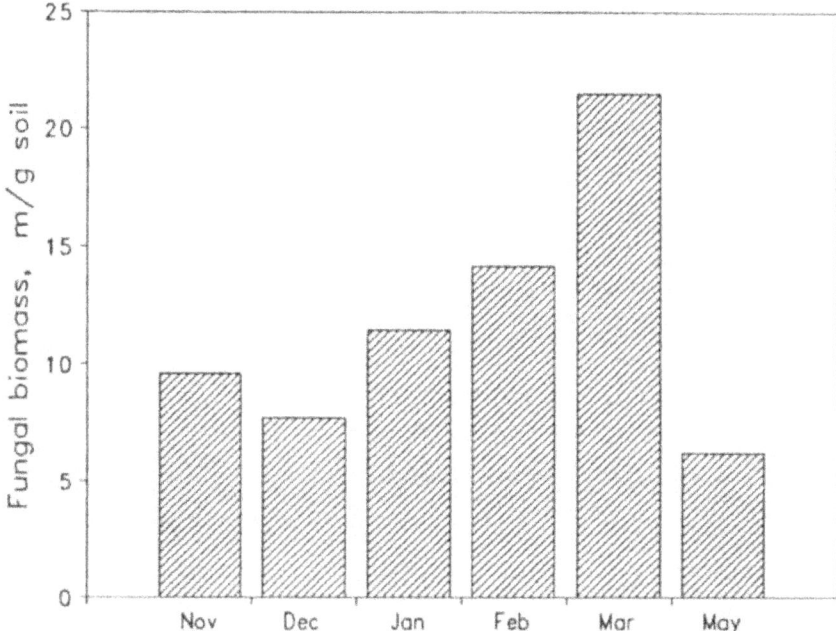

Figure 4.20 *Changes in fungal biomass in top soil in wet miombo at Misamfu in northern Zambia*

In order to determine nutrient loss during decomposition, subsamples of litter should be analysed for nutrient content at the beginning and end of the decomposition period (Table 1.10). Such analyses are useful for understanding nutrient cycling in the ecosystem.

There is a seasonality in the rate of decomposition in miombo which is related to both rainfall and soil microbial activity (Malaisse *et al.* 1975). In northern wet miombo soil fungal biomass reaches a peak at the end of the rainy season (Figure 4.20). For this reason, it is also important to assess populations of soil microbes. The reader is referred to more specialized literature on techniques for assessing microbes (e.g. Swift *et al.* 1979; Jackson and Raw 1966).

Soil macrofauna also play an important role in litter decomposition and nutrient cycling in miombo. The miombo soil macrofauna is very diverse and diversity varies spatially. Among the abundant groups are insects, collembola, mites and millipedes. Termites made up 80 per cent of the total soil fauna biomass in a Zairean wet miombo (Malaisse *et al.* 1975) and although these were numerically dominant in a Zimbabwean dry miombo, millipedes and beetle larvae accounted for 72 per cent of the fauna biomass (Dangerfield 1993). Soil macrofauna can be assessed by hand sorting visually or with the aid of a magnifying glass in soil monoliths. Dangerfield (1993) used this method to assess the abundance of macrofauna in 25 × 25 × 30 cm soil monolith samples. Their overall abundance in undisturbed miombo was 197 individuals m^{-2}, with a biomass of 12.5 g. The fauna at a cleared miombo site was more than double that in the adjacent undisturbed miombo site.

CHAPTER 5
Management Guidelines

FOREST MANAGEMENT is concerned with the maintenance and regeneration of forest products and ecosystem services. Maintenance can be achieved through the proper management of old-growth forests, while regeneration can be achieved through natural recovery and plantation silviculture.

Natural regeneration refers to the renewal of a tree crop by self-sown seed or vegetative regrowth with or without silvicultural practices, while plantation regeneration refers to the use of standard practices of plantation silviculture to create a tree crop from sowing seeds or planting seedlings. Plantation regeneration of indigenous trees in the Zambezian phytoregion has depended largely on standard nursery practices inherited from commercial forestry of exotic species (Piearce 1993). This may have contributed to the poor success of miombo tree planting. In many cases plantation silviculture of indigenous trees has not taken advantage of the natural regenerative characteristics of miombo woodland (Appendix 11). Although it has been known for a long time that the natural regeneration of miombo is easier and cheaper, official policies have ignored this and often opted for plantation silviculture which is better suited for fast-growing exotic trees such as *Eucalyptus* and tropical pines. However, both the natural and artificial regeneration of indigenous trees have a role to play in indigenous forest management, depending on species characteristics, local environment and socio-economic conditions.

Table 5.1 *Stem sizes of some miombo trees at the age of 10 years in coppiced regrowth and plantation plots in the Zambian Copperbelt*

Species	Mean girth (cm) at 1.3 m above ground	
	Coppiced regrowth	Plantation
Brachystegia utilis	13.0	<1.0
Julbernardia paniculata	16.8	<1.0
Parinari curatellifolia	19.7	30.8
Pterocarpus angolensis	17.7	12.9
Strychnos cocculoides	9.3	25.5
Uapaca kirkiana	14.0	12.9

Based on Chidumayo (unpublished) for coppiced regrowth data and Sekeli (1993) for plantation data

The differential performance of miombo trees under natural and plantation conditions shows the importance of species characteristics (Table 5.1). For example, *Brachystegia* and *Julbernardia* species are clearly unsuitable for plantation regeneration on account of their very slow growth rate from seed, but these species perform better under natural regeneration as coppiced regrowth. Without considering costs, the fruit trees *Parinari curatellifolia* and *Strychnos cocculoides* perform better under plantation regeneration, while *Pterocarpus angolensis* and *Uapaca kirkiana* seem to perform equally well under both types of regeneration.

Natural regeneration

Woody plant density in old-growth miombo is dominated by stunted old seedlings (Table 1.6). Since many of these seedlings have well-developed roots, their failure to grow into the sapling phase appears to be caused by canopy shading by older trees. The creation of gaps in the canopy should therefore initiate patch regeneration which results in an uneven age stand. Gaps are naturally created through the death of canopy trees or the loss of branches through wind and lightning. Artificial gaps in the canopy are created by tree harvesting. The characteristics of natural regeneration in miombo are therefore likely to be influenced by the type of harvesting.

Harvesting techniques
Miombo has been harvested by either selective or clear cutting. Selective harvesting involves the cutting of trees of a particular size range or species, while clear cutting either involves the cutting of all trees or the majority of trees with a few left uncut. Selective cutting is the predominant form of harvesting for specific wood products such as poles and timber, as well as for edible caterpillars which show preferences for certain food trees. Because selective harvesting creates irregular gaps in the canopy, natural regeneration is also irregular and patchy, and the resultant age and size structure of the stand is uneven. In some cases the reduced competition caused by selective cutting may enhance the growth of the uncut trees, which may suppress the regeneration of stunted seedlings in the herb layer. If this occurs, selected species may fail to regenerate. This may be one of the reasons why the regeneration of timber species such as *Pterocarpus angolensis* and *Baikiaea plurijuga* has not been very successful following selective cutting. In miombo where the proportion of valuable timber species is small, selective cutting may further endanger the regeneration of such species.

The following guideline is suggested for selective cutting in miombo. Selective cutting should be based on species, and should be carefully planned and implemented to ensure that the exploited species does regenerate. This can be done by either clear cutting the stand following selective cutting, or cutting neighbouring trees to create larger canopy gaps that will free seedlings from shading by trees normally left uncut during selective cutting.

Table 5.2 *The species diversity in uneven age old growth and coppiced regrowth miombo in Zambia*

Miombo subtype	Mean number of species 0.1 ha⁻¹	
	Old-growth	Regrowth
Northern wet miombo	18.30	24.20
North-western wet miombo	18.78	24.12
Central dry miombo	14.79	13.28
Eastern dry miombo	16.65	20.60

Based on Chidumayo (1987)

Clear cutting removes the entire canopy with or without reserved trees. In miombo this is the best way of encouraging regeneration from the stump coppice and from seedlings. Such regrowth stands usually have higher species diversity than the old-growth stands which they replace (Table 5.2), and the danger of species loss is greatly minimized.

The main disadvantage of clear cutting is its negative effects on catchment hydrology (Figure 4.18). In order to minimize the negative effects, strips or coupes to be cleared should alternate with shelterbelts (Figure 5.1). The Forest Department in Zambia has used this technique in forest reserves that have been harvested for woodfuel. The Department demarcates the coupes prior to allocation to woodfuel harvesters and these coupes are usually wider than the shelterbelts (Figure 5.1). Although the initial purpose of the shelterbelts was to serve as seed sources for regeneration in the cleared coupes, given the pattern of regeneration in miombo (see Chapter 1), these are probably more important for minimizing hydrological disturbances to catchment areas (Serenje *et al.* 1994). Once adequate regrowth has occurred in the coupes, which is usually after 10 years, the shelterbelts may be cleared, with the regrowth strips performing the role of shelterbelts. The effectiveness of shelterbelts is enhanced if they are oriented across the slope.

Trees are cut in different ways. In the chitemene and other forms of shifting cultivation, trees are either pollarded by de-branching or lopping, with the objective of hastening the regeneration of biomass and the canopy. The regrowth stems or branches are invariably defective at the point of cutting. Consequently the pole and timber value of such regrowth is low, although the regrowth is useful as a natural fallow and as a source of nutrients for crop production. Pollarding and lopping are therefore not recommended if the objective of regeneration is to produce poles and timber.

Tree cutting for woodfuel, poles and timber is done by stumping (cutting at base or close to the ground). In Zambia the Forest Department requires that stumps are less than 30 cm AG and that these are cut at an angle to prevent water collection and stump rot. Land clearing for semi-permanent and permanent cultivation involves stumping or uprooting trees. Miombo

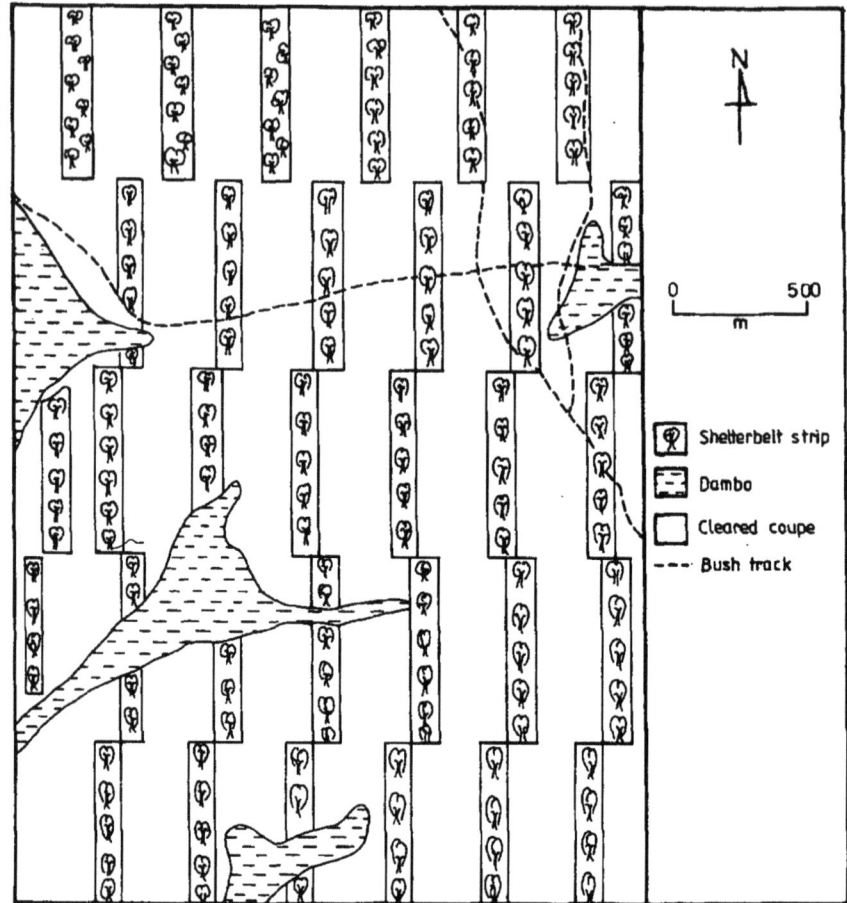

Figure 5.1 *Clear-cutting with shelterbelts in miombo woodlands (From Serenje et al. 1994)*

regeneration in stumped and uprooted areas, including those abandoned after cultivation, is predominantly of root origin. Unless damaged by fire (see below), such regeneration produces dense stands of straight poles and can be managed for the production of poles in the intermediate term, and for timber in the long term. Because old-growth miombo stands are dominated by defective stems resulting from past mistakes (Hursh 1960), the improvement of the pole and timber value of such defective stands requires clear cutting by stumping or uprooting to encourage the production of poles and timber stems.

Clear cutting with a few reserved trees has the same effect as complete clear cutting. The uncut trees are reserved for specific purposes. The reserved trees are either legally protected timber species (*Afzelia quanzensis, Baikiaea plurijuga, Entandrophragma delevoyi, Faurea saligna, Khaya nyasica, Pterocarpus angolensis*) or fruit species (*Anisophyllea boehmii, Parinari curatellifolia, Strychnos cocculoides, Uapaca kirkiana*). In many traditional farming systems the reserved trees serve many functions and belong to a variety of species (Table 5.3). In Zimbabwe, trees are left when clearing land for fruit, shade, fertility from leaf litter and for spiritual reasons (Scoones *et al.* 1993). Coppice with standards is the resultant regrowth in areas clear cut with reserved trees. The regrowth has the same pole and timber qualities as that from complete clear cutting by stumping and uprooting, but the reserved trees provide products that would not be available immediately in coppice regrowth. Clear cutting with reserved trees should therefore be encouraged by forest and agriculture extension workers.

The time when trees are cut affects coppicing vigour and production during the first year. Chidumayo (1993a) found that cutting in July and November resulted in lower productivity than cutting in October (Table 5.4). Both seasonality in climatic factors, especially temperature, and phenology appear to affect coppicing vigour and the productivity of sprouts after tree cutting. Cutting in September and October probably results in the most vigourous and productive coppice in Zambian miombo woodland.

Cutting cycles

Cutting cycles for regrowth miombo depend on the purpose of management. Given that natural regeneration and productivity are highly variable in space and time, only broad cutting cycles can be suggested based on the limited

Table 5.3 *The structure of reserved trees in cultivated plots in three miombo woodland areas in Zambia*

Variable	Mongu	Ndola Rural	Solwezi
Mean plot size (ha)	2.39	1.68	1.20
Number of preserved species	16	32	18
Preserved fruit species/plot	7	10	6
Mean preserved species/plot	4.0	2.7	1.4
Mean preserved tree/plot	4.6	2.9	1.7

After Chidumayo and Siwela (1988)

Table 5.4 *The effect of the timing of tree cutting on production during the first growing season.*

	Coppice production during the first growing season (g m^{-2})		
Cutting time	Leaf	Wood	Total
July	13.0	17.5	30.5
October	22.4	25.5	47.9
November	7.7	9.4	17.1

Based on Chidumayo (1993a)

Table 5.5 *Suggested cutting cycles for natural regeneration in miombo*

Forest product	Size/quantity	Rotation period (years)
Poles:		
small	12–25 cm gbh[1]	10–20
medium	26–38 cm gbh	21–30
large	39–63 cm gbh	31–50
Sawn timber	>63 cm gbh	>50
Above ground wood biomass:		
shifting cultivation	25–35 t ha^{-1}	13–19
firewood (pole size)	12–63 cm gbh	10–50
charcoal (large poles)	39–63 cm gbh	31–50

1. gbh is girth at 1.3 m above ground

Table 5.6 *Mortality among marked trees at Ndola old-growth miombo plots maintained under different fire treatments for 11 years*

	Annual tree mortality (%)
Fire protected plot	0.38
Early burnt plot	0.64
Late burnt plot	1.58

Based on Trapnell (1959)

available production data. Above ground biomass production in regrowth miombo has been estimated at 2.7 t and 1.9 t ha^{-1} yr^{-1} in wet and dry miombo, respectively (see Chapter 1). From the available data, stem gbh increment has been estimated at 1.25 cm yr^{-1} in both dry and wet miombo. On the basis of these production figures, cutting cycles for the different wood products in regrowth miombo have been suggested in Table 5.5.

Of course, given the very high stem densities in regrowth miombo, small wood products (small poles) can be harvested as part of a thinning silvicultural practice, and the remaining stems managed for large wood products (large poles, sawn timber and cord wood for charcoal). For example, regrowth can be thinned after 10–20 years by the selective harvesting of small poles, and the stems of timber trees can be reserved for timber production and harvested after 50 years or a longer period. This improvement in cutting practice might even enhance growth in the remaining stems.

Fire management

Fire is probably the most important management problem in miombo which must be resolved before any forestry management can be implemented. All fires kill the shoots of seedlings and young regeneration. But although complete exclusion from young stands is necessary, it may not be possible. The fiercest and therefore most damaging fires occur during the late dry season because of the favourable fire conditions: high quantities of extremely dry litter biomass (see Chapter 3). In recently cut-over areas, the amount of wood debris is usually high and even an early dry season fire can be fierce and damaging to young regeneration. Fire management in miombo should therefore take into account the age and size of the woodland, the phenology of the dominant or desirable species and the type of land use.

In old-growth miombo little can be achieved by fire control, although continuous late dry season burning increases tree mortality (Table 5.6). However, it is rare under the traditional burning regime for a site to be continuously late burnt. Irregular burning is therefore the most economical silvicultural practice in old-growth miombo.

The productivity and quality of stems in regrowth miombo can be reduced by fire. Not only is wood production reduced, but the mean annual increment is also reduced (Figure 5.2) due to wood destruction by fire, especially late dry season fire. Fire is therefore a principal agent in preventing regeneration and crippling young stems. This crippling effect takes place in the early stages of regeneration. Crippling fires also cause basal scars and stem lesions in older stems, which produce defective stems of poor pole and timber quality.

In regenerating areas, fire management should involve the following:

1. At the time of cutting all discarded wood should be piled away from stumps and patches of dense saplings and later burnt. Piling wood around stump bases is a common practice in Zambia when clearing land for cultivation. However, in areas managed for forest production the practice of piling wood debris anywhere without regard to stumps and advanced sapling regeneration should be discouraged by extension workers.
2. Young regrowth areas should be protected from late dry season fires by early dry season burning which should be continued until leaf production has reached the level of old-growth miombo. This usually occurs when the regrowth is 10–15 years.
3. Older regrowth areas should be regularly burnt annually before leaf flush (September) to minimize damage to new leaves and reduce the build-up of litter biomass that occurs when areas are not burnt.

Traditional tree conservation practices
The most common traditional practice in the miombo region is to selectively retain some trees in fields and villages at the time of clearing (Box 5.1). However, farmers establishing new fields can leave shelterbelts or boundary strips of indigenous woodland of variable width for soil and water conservation. In Zambia, woodland within 30 m of a stream or river is protected by law and should be reserved to prevent bank erosion.

Where the woodland was cleared, it is possible to encourage natural regeneration from stumps and roots on contour bands and boundary strips

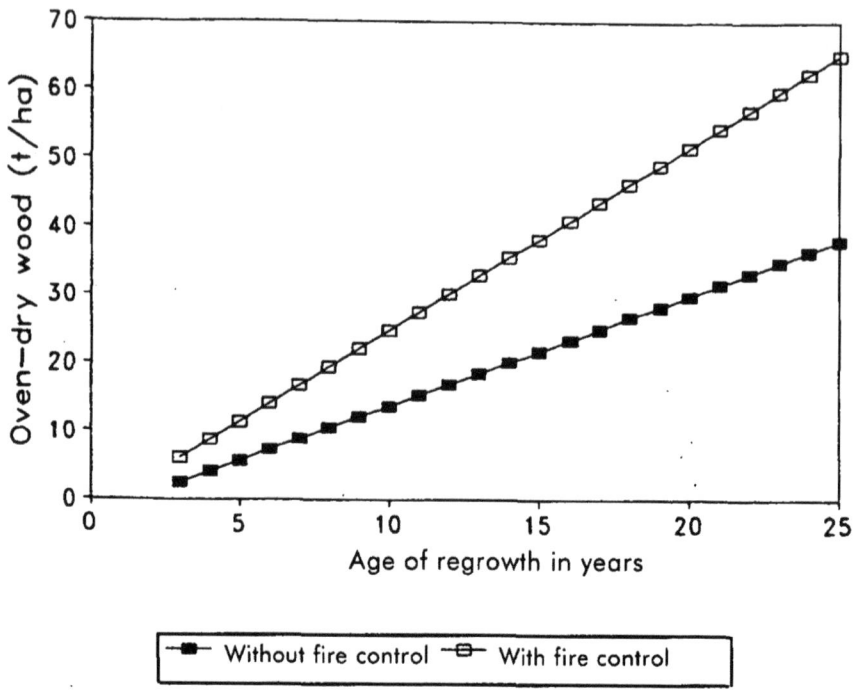

Figure 5.2 *The effect of fire control on above ground wood biomass accumulation in regrowth dry miombo*

by allowing coppice to develop. The regrowth can then be managed by thinning to harvest small wood products.

Seedlings from naturally sown seed in cultivated fields can also be nurtured for the production of specific products. In the Kafue flats area of southern Zambia, farmers deliberately nurture self-sown seedlings of *Faidherbia albida* in fields and around cattle kraals by encouraging the development of straight vigorous stems through periodic pruning for 3–5 years until the remaining lateral branches are high enough (Olsen 1992). Such techniques can easily be applied to many miombo trees.

Grazing and browsing

In old-growth miombo, canopy shading depresses herbage production which is required for grazing. Opening up the canopy by thinning or clear cutting promotes herbage production (Hood 1972) and therefore improves grazing. Tree cutting should be done by lopping at about breast height so that trunk coppice is accessible to browsers but high enough to minimize damage by fire. Browsers tend to over-exploit preferred species which may be killed (Figure 5.3). It may be necessary therefore to carry out rotational grazing to prevent mortality of preferred browse species.

Box 5.1 *Tree conservation in fields and villages in Zambia*

It is quite common in Zambia to find both indigenous and exotic trees in fields and around homesteads. Exotic trees are predominantly fruit trees (Chidumayo 1988c) raised from seed and seedlings. The majority of indigenous trees are left when clearing land for building villages and cultivation. Very few are actually raised from seed and seedlings. The reasons for conserving indigenous trees vary with locality as shown in the table below. The common reason for leaving trees in fields is that they are either too hard to cut or they produce edible fruit, while in homesteads trees are commonly left for shade and fruit.

	Common reason for conserving trees	
Species	In villages	In fields
Pericopsis angolensis	Shade	Too hard to cut
Brachystegia spp.	Shade	Too hard to cut
Parinari curatellifolia	Fruit	Fruit
Uapaca kirkiana	Fruit	Fruit
Syzygium guineense	Fruit	–
Pterocarpus angolensis	Shade	Too hard to cut
Julbernardia spp.	Shade	Too hard to cut
Swartzia madagascariensis	Shade/Medicine	Too hard to cut
Anisophyllea spp.	Fruit	–
Monotes spp.	Shade	–
Isoberlinia angolensis	Shade	–
Vangueriopsis lanciflora	Shade	Fruit
Strychnos cocculoides/S. pungens	Fruit	Fruit
Afzelia quanzensis	Shade	–
Piliostigma thonningi	Shade	–
Marquensia macroura	Shade	–
Diplorhynchus condylocarpon	Shade	–
Erythrophleum africanum	Shade	Too hard to cut
Ficus spp.	Shade	
Albizia antunesiana	–	Too hard to cut
Combretum molle/C. psidiodes	–	Too hard to cut
Faurea spp.	–	Too hard to cut

Based on survey carried out in 1986 in Serenje, Mongu; Solwezi and Ndola Rural districts (Chidumayo unpublished)

In parts of Luapula and Northern Provinces, small patches of indigenous woodland, usually < 0.10 ha, are maintained near homesteads as areas for chickens to forage which also offer protection from birds of prey. In some cases forest products such as small poles and medicines are harvested from such natural woodlots.

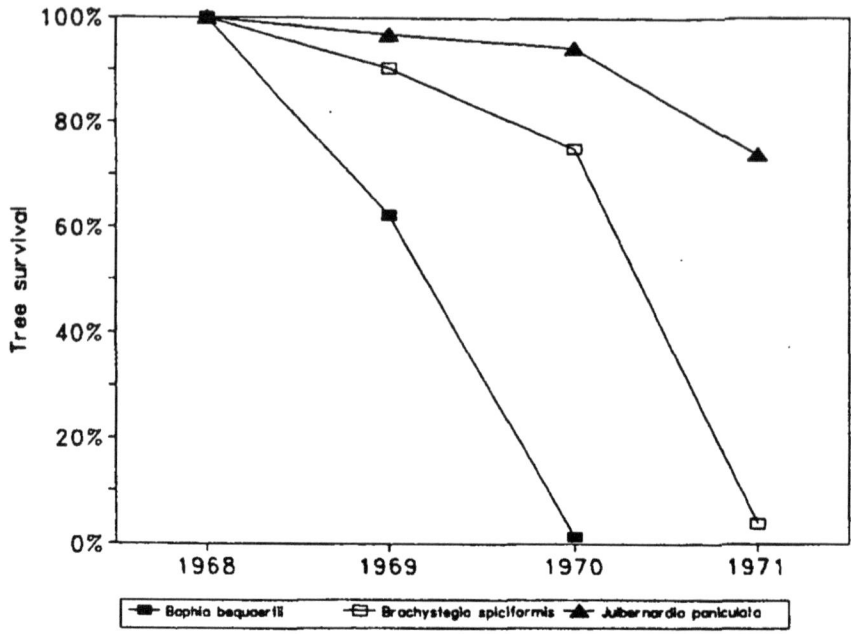

Figure 5.3 *The mortality of coppiced tree stumps caused by overbrowsing by cattle in experimental wet miombo paddocks (Based on Hood 1972)*

Bush encroachment species tend to be fire-intolerant and usually flourish better where fire has been excluded either through management or grazing. Grazing areas with a bush encroachment problem require late dry season burning once in several years to control encroachment. This management practice is also used to control ticks and insects causing livestock disease, whose populations may build up in the absence of fire.

Plantation regeneration

Species selection
The selection of a species for planting depends on the purpose it serves and the products it yields. Trees are planted for a variety of reasons, among which are to produce poles and timber, fruits, to improve soil fertility, to provide shade and to produce live fences. In a few cases trees may be planted for medicinal purposes. The majority of past attempts to plant miombo trees have aimed at providing timber or fruits.

The selection of plantation trees for timber is primarily determined by growth rate and form. Plantation trials of miombo trees have shown a range of growth rates (Figure 5.4 and Figure 5.5). What has emerged from these trials is that the dominant miombo species (*Brachystegia, Isoberlinia* and *Julbernardia*) are very slow growing when raised from seed, while a

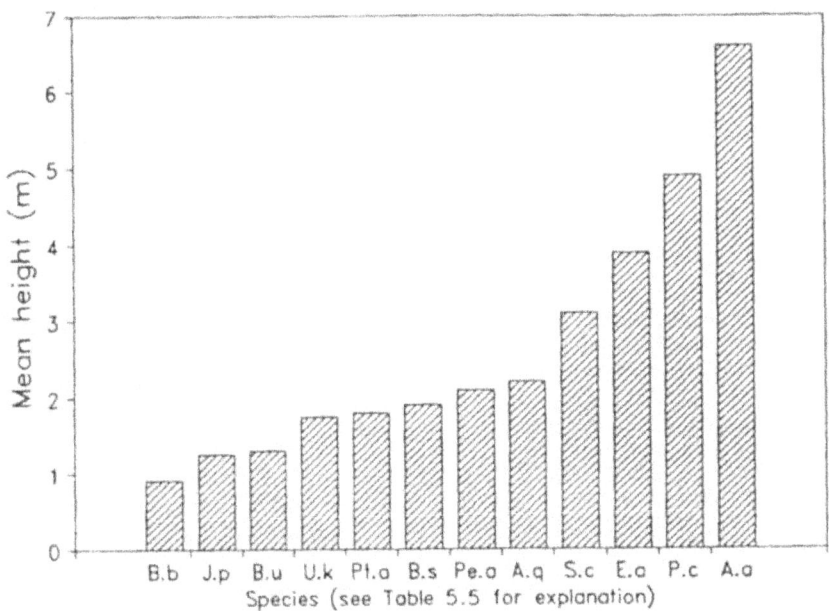

Figure 5.4 *Mean stem heights of ten-year old miombo trees under plantation conditions in Copperbelt Province, Zambia*

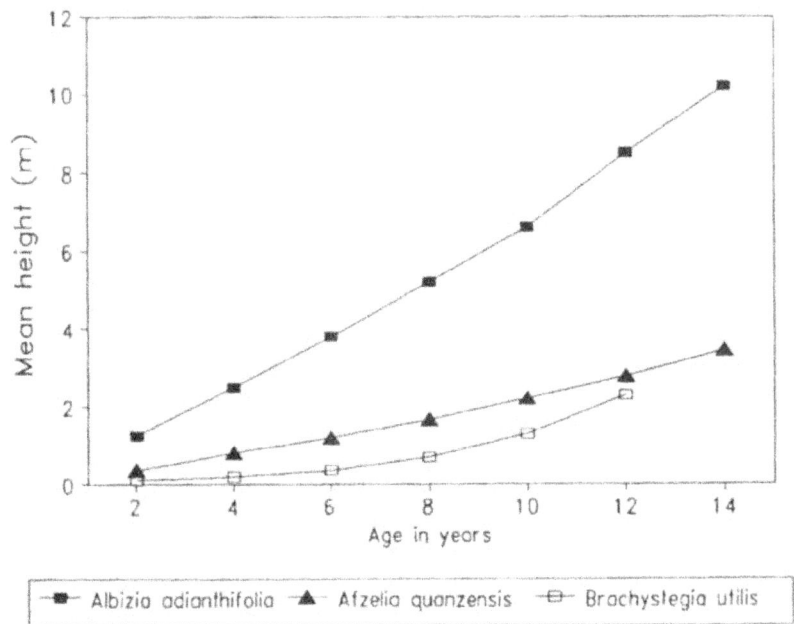

Figure 5.5 *Stem height growth of three miombo trees under plantation conditions in Copperbelt Province, Zambia*

Table 5.7 *Some performance criteria to guide the selection of miombo trees for plantation regeneration*

Species	Maximum seed germination rate %	rank	Recorded seedling survival rate after one year %	rank	Mean dbh in plantation trials after ten years cm	rank
Afzelia quanzensis (A.q)	99	1	88	2	3.9	6
Albizia adianthifolia (A.a)	30	9	90	1	10.1	1
Brachystegia boehmii (B.b)	69	5	na	–	<1.0	7
Brachystegia spiciformis (B.s)	83	3	78	5	<1.0	7
Brachystegia utilis (B.u)	79	4	na	–	<1.0	7
Erythrophleum africanum (E.a)	10	10	na	–	6.3	4
Julbernardia paniculata (J.p)	95	2	45	6	<1.0	7
Parinari curatellifolia (P.c)	10	10	80	4	9.8	2
Pericopsis angolensis (Pe.a)	40	8	90	1	3.9	6
Pterocarpus angolensis (Pt.a)	60	6	85	3	4.1	5
Strychnos cocculoides (S.c)	60	6	90	1	8.1	3
Uapaca kirkiana (U.k)	80	7	20	7	4.1	5

Notes: dbh is diameter at 1.3 m above ground and na denotes data not available. Based pm Chidumayo (1993a) and Sekeli (1993)

Table 5.8 *Vegetation conservation in forest and national parks in Zambia*

	Extent in million hectares			
	Protected areas		Unprotected areas	Total
Vegetation	Forest reserves	National parks		
Forest:	0.63	0.30	3.75	4.68
Evergreen	0.35	0.23	2.86	3.44
Deciduous	0.23	0.07	0.57	0.87
Montane	0.00	0.00	0.08	0.08
Swamp	0.00	0.00	0.15	0.15
Riparian	0.00	0.00	0.09	0.09
Plantation	0.05	0.00	0.00	0.05
Miombo woodland:	5.47	3.10	30.82	39.39
Wet	3.41	0.60	18.42	19.37
Dry	1.23	1.80	9.56	11.48
Kalahari	0.83	0.70	7.01	8.54
Savanna woodland:	0.40	1.33	7.49	9.22
Mopane	0.23	0.99	2.63	3.85
Munga	0.12	0.11	2.72	2.95
Termitaria	0.05	0.23	2.14	2.42
Grassland:	0.94	1.21	18.53	20.68
Wetland	0.01	0.54	12.46	13.01
Dambo	0.93	0.67	6.07	7.67
Aquatic:	0.00	0.00	1.29	1.29
Natural	0.00	0.00	0.66	0.66
Man-made	0.00	0.00	0.63	0.63
Total	7.44	5.94	61.88	75.26

Based on 1:500,000 vegetation map of Zambia (Edmonds, 1976; Schultz, 1974; Adeyoju, 1991 and Chidumayo, 1994a) few species such as those of *Albizia, Parinari* and *Erythrophleum* are relatively fast growing with potential for plantation silviculture. The valuable timber species, such as *Pterocarpus angolensis* and *Afzelia quanzensis*, have intermediate growth rates and perhaps limited potential for plantation silviculture.

Plantation trials of fruit trees have been restricted to a few species, such as *Strychnos cocculoides* and *Uapaca kirkiana*. The latter species is slow growing while *S. cocculoides* is relatively fast growing and is reported to fruit at the age of 6–7 years (Sekeli 1993). The potential for growing these trees, especially on a large scale, is therefore limited.

Other factors that determine the suitability of trees for plantation silviculture include seed availability and germination rate, and seedling survival rate. Table 5.7 is an attempt to rank some miombo trees on the basis of these criteria. Apparently species that are relatively fast growing (*A. adianthifolia*, see Figure 5.5) have low seed germination rates. Seed treatment to enhance germination may therefore be necessary for these species. Except for *U. kirkiana*, seedling survival rates of miombo trees appear to be good. However, seedling die-back (see Chapter 1) may discourage the growing of these trees. With the limited experience in plantation silviculture of miombo trees, careful species selection is necessary. Disease can be a problem even in indigenous trees. Blight caused by the soil-borne *Fusarium oxysporum* fungus has been reported to kill *Pterocarpus angolensis, Burkea africana, Erythrophleum africanum, Lannea stuhlmanii, Strychnos cocculoides* and *Terminalia sericea* (Geary 1972). Mortality among *P. angolensis* may be as high as 36 per cent and in almost pure stands infestation rates reach 60 per cent (Bainbridge and Edmonds undated). Serious infestation of *Julbernardia globiflora* and *Brachystegia spiciformis* by an *Aspidoproctus* scale insect has been reported in Zimbabwe, while a bark-browsing carpenter moth, *Salagena* sp., adversely affected *Baikiaea plurijuga* trial plantations in Zambia (Piearce 1993). Wild fruit trees are hosts to a variety of pests which affect fruit production either directly through infestations or indirectly by attacking foliage and shoots which interferes with photosynthetic efficiency. Boring weevils and beetles cause premature fruit drop in *Parinari curatellifolia* and *Uapaca* species (Mwamba 1989). Disease control will therefore become an important aspect of plantation silviculture of indigenous trees.

Seed collection and storage
Raising indigenous trees from seed is limited by seed availability, storage problems and pre-treatment requirements (Piearce 1993), because sowing seeds and raising seedlings require seed collection, storage and sometimes pre-sowing treatment. Seed yield in miombo trees is unreliable due to erratic and irregular fruiting, and seed predation by insects. Seeds lose much of their viability if they are not handled properly and in some species moisture content affects viability. Low moisture content (< 15 per cent)

drastically reduces seed viability in *Isoberlinia angolensis* (Chidumayo 1993a) and has also been reported to reduce the viability of *Uapaca kirkiana, Anisophyllea boehmii* and *A. pomifera* seeds (Mwamba 1989).

The following should therefore be considered when collecting seeds for plantation regeneration:

(1) Seeds should be collected from ripe fruits or pods or on the ground immediately after dispersal and the right period to collect seeds in miombo is from August to December when the fruits of the majority of trees ripen (see Figure 1.6).
(2) Seeds should be collected from trees with the desired characteristics, for example, those that are straight, vigorous and healthy.
(3) To ensure adequate genetic variation, seeds should be collected from many trees, and since miombo trees show annual variation in fruit production (see Figure 1.15), seeds should be collected in years when there is a high frequency of fruiting trees.
(4) Seeds in fruits or pods should be extracted by drying and threshing. Some pods, such as those of *Pterocarpus angolensis*, require roasting to burn off the spikes, or soaking in hot water to soften the pods before opening with a sharp knife. Pulpy fruits can be cleaned by soaking in water for a day or two before removing the pulp.
(5) After extraction, the seeds should be sorted to remove dirt, chaff and damaged seeds by sieving or floating in water. Non-viable seeds normally float in water and can therefore be easily separated from viable seeds.
(6) If the seeds are not immediately required for sowing, they should be stored in containers that are tightly sealed to keep away predators, especially insects. The species, provenance and date of collection should be clearly labelled on seed containers and for long- term storage, seeds should be stored in plastic containers to avoid the uptake of water vapour and kept in a cool place (0–20°C).

Seed sowing

Seeds of the majority of miombo trees have no dormancy or resting stage before germination and therefore do not require pre-sowing treatment. However, germination rate of some species such as *Acacia polyacantha, Albizia* spp., *Amblygonocarpus andongensis, Dichrostachys cinerea, Erythrophleum africanum, Pericopsis angolensis* and *Swartzia madagascariensis* is low (<20%) but can be raised by pre-sowing seed treatment. Such treatment includes placing seeds in hot (but not boiling) water for 24 hours or in cold sulphuric acid for a few minutes and rinsing the seeds in running water for up to 24 hours. However, working with acid can be dangerous and acid may not be easily available in some areas. Seeds can also be mechanically treated by scarring the testa with a file or sharp

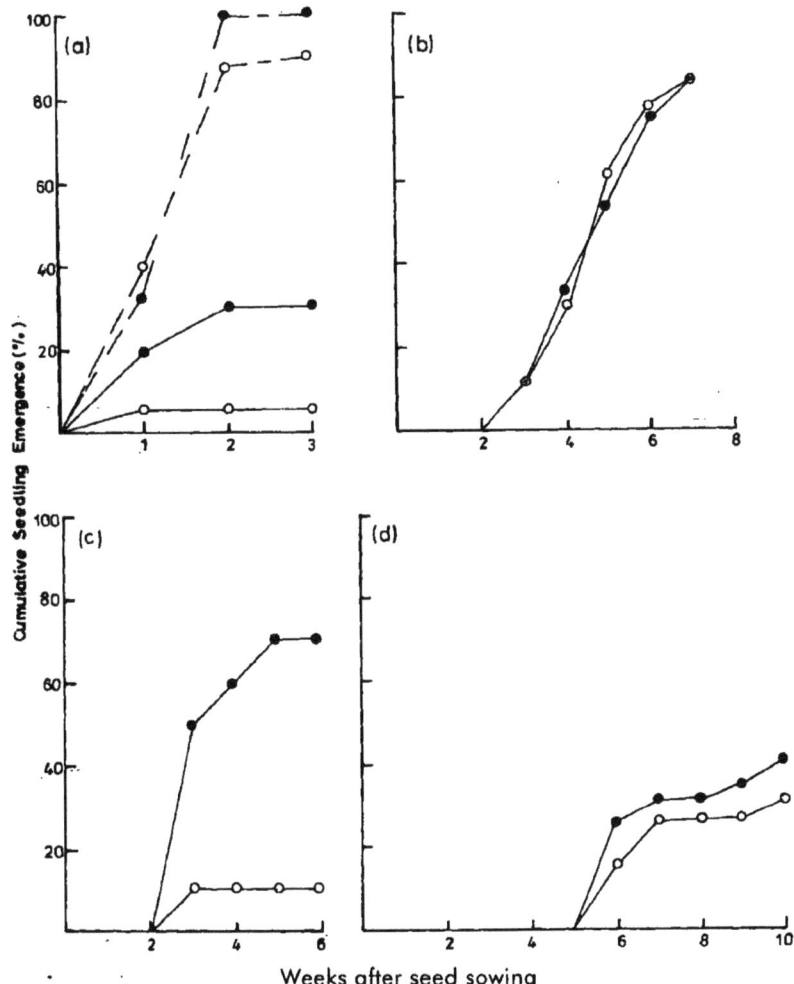

Figure 5.6 *Seedling emergence of (a)* Acacia polycantha, *(b)* Afzelia quanzensis, *(c)* Tamarindus indica, *(d)* Bauhinia petersiana *in untreated soil (open circles) and earth-kiln burnt soil (solid circles). Broken lines in (a) represent the use of scarred seeds. Other seeds were untreated.*

knife. Sowing seeds in soil that has been pre-heated by burning wood also improves germination (Figure 5.6). The germination of seeds encased in hard shells, such as those of *Parinari curatellifolia*, can be improved by cracking the case and sowing the naked seed.

Nursery techniques
Because of the limitations of plantation silvicultural research and knowledge of indigenous trees, the majority of nursery techniques are based on

commercial forestry with exotic species. Further research is therefore necessary to develop appropriate nursery techniques for indigenous trees. Nursery techniques aim at the successful production of plantation seedlings that are healthy and vigorous to ensure high survival rates in the field at minimum cost. The following guidelines are suggested.

Nursery seedlings should be raised in tall pots of polythene or other available local materials which should be filled with miombo soil. Seedlings should be transplanted when roots start to outgrow the pot or when the shoot is 10–40 cm. Although root pruning promotes lateral root production, it has been observed to depress seedling growth in *Pterocarpus angolensis* and *Julbernardia globiflora* (Munyanziza 1994) and should therefore be avoided when raising miombo tree seedlings in the nursery. Miombo soil is generally poor in nutrients and an application of NPK fertilizer may be required to promote seedling growth. However, the application of NP fertilizer increases shoot growth at the expense of root growth (Munyanziza 1994). During the establishment period miombo tree seedlings allocate more biomass to roots, and the application of fertilizer reverses this natural growth strategy and should therefore be undertaken with great care.

Miombo trees have symbiotic relationships and the most important of these are root-nodule bacterial and mycorrhizal partnerships that enhance plant nutrition (Högberg 1992). The ecto-mycorrhizal habit is predominant in non-nodulating genera such as *Brachystegia, Isoberlinia* and *Julbernardia*. Plantation silviculture of miombo trees should therefore consider the role of these symbiotic relationships. For example, it has been shown that wildlings of *Uapaca kirkiana* naturally infected by fungi, and nursery seedlings inoculated with appropriate fungi, have higher survival and growth rates than seedlings that have not been inoculated (Mwamba *et al.* 1992, Mwamba in press). *Brachystegia microphylla* seedlings inoculated with fungi also show significantly higher growth rates than non-inoculated ones (Munyanziza 1994).

Direct sowing in the field involves sowing the seed on the soil surface or 1–2 cm into the soil. Direct sowing in the field is recommended for the majority of miombo trees because they develop a very deep tap root before there is any significant shoot growth. However, raising seedlings in the nursery is recommended when seeds are scarce or very small. In some species, such as *S. madagascariensis* and *U. kirkiana*, root development is very slow and may require a longer nursery life before outplanting in the field.

Wildlings that regenerate naturally from dispersed seeds can be carefully uprooted with soil around the roots and transplanted. Only the current year's seedlings should be used as older ones do not survive transplantation, and this should be done early to allow for establishment in the field before the end of the rainy season.

Vegetative propagation
A few miombo trees, such as *Pterocarpus angolensis* and *Lannea discolor*, can be raised from cuttings of various sizes. A cutting is a piece or section of a stem, branch, twig or root from a parent tree. The cutting is placed in the soil upright, either directly in the field or in a nursery. Cuttings should be 30–50 cm long if they are planted directly in the field with sufficient internodes in the ground to permit adequate rooting. The best cuttings come from young branches (1–2 cm diameter) of healthy parent trees. Large and very long cuttings are often used for live fencing, but the mortality rate can be very high because such cuttings do not develop an adequate root system to support the large AG biomass.

Planting and tending
Nursery plants should be watered before transportation and potted plants should be transported to planting sites in wooden trays for easy handling.

Trees should be planted in straight rows for easy tending and at a higher density to allow for thinning by natural mortality and for some harvesting.

Seedlings should be retained in polythene pots when planting out to ensure a better start in the field. However, in some cases polythene pots may distort root growth and should therefore be removed when planting out.

Weeding by machine or hand is necessary to reduce competition and fire hazard and this treatment should be continued until seedlings have successfully established. Fast- growing species require shorter weeding periods (2–5 years) than slow- growing species (6–10 years).

The majority of indigenous trees are not damaged by termites, but many are susceptible to fungal infestations (Piearce 1993). Prophylactic or curative treatment with fungicides may therefore be necessary.

Enrichment planting on charcoal spots
Wood carbonization at earth kiln sites results in the destruction of root stocks of woody plants. Because of this tree regeneration on charcoal spots is impaired for 10–20 years (Chidumayo 1988b). Apparently seedling establishment is made difficult by altered edaphic conditions (see Chapter 4). Fires are also fierce on charcoal spots due to high herbage production. In the past, the Forest Department has sown seeds of exotic trees such as *Eucalyptus* and *Gmelina* on charcoal and ash spots as enrichment planting (Chidumayo 1993d); but this silvicultural practice was discontinued following establishment failure caused by drought and fire (Lees 1962). Direct seed sowing with suitable indigenous trees that are tolerant of neutral soils at charcoal spots and with early deep-rooting habit might produce better results. Chidumayo (1994c) has shown that *Acacia polyacantha* performs better in charcoal soil than miombo trees under laboratory conditions. Further field research is required to identify species that can perform well on charcoal spots.

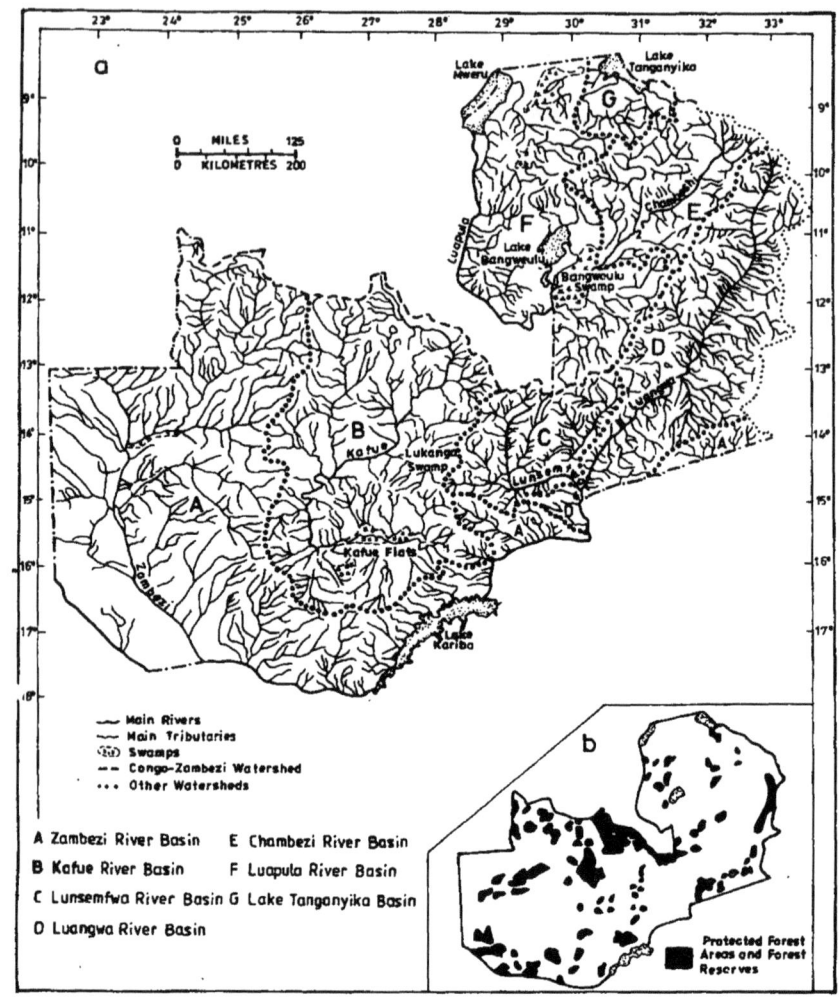

Figure 5.7 *River basins (a) and forest reserves (b) in Zambia*

Catchment and biodiversity conservation

Water balances in catchment areas can easily be upset by disturbances such as deforestation (see Figure 4.18). Watershed catchments should therefore be maintained under as natural conditions of old-growth woodland as is possible. This is particularly important for vulnerable areas such as hills, escarpments and the headwaters of streams and rivers.

The management of watersheds in Zambia has been implemented through the creation of forest reserves. However, the size and distribution of forest reserves is currently inadequate (Figure 5.7). Larger areas in the major river headwaters are required to protect water and soil resources in river basins effectively.

Forest reserves (including botanical reserves) and national parks (Figure 2.9 and Figure 5.7) were also established for ecosystem and biodiversity conservation. Miombo is the dominant vegetation in forest reserves and national parks in Zambia (Table 5.8).

Miombo is a high species-diverse vegetation in which adequate biodiversity conservation depends on the size of the conservation areas. Conservation areas of 1000–2500 km^2 have been suggested for tropical rainforests, while 500 km^2 is probably the minimum for most forest types (Lebec and Goodland 1988). If conservation aims at rare species which occur in low densities, then large areas will be required to support viable breeding populations. The low species dominance and high species diversity in miombo suggests that large areas are required for the conservation of biodiversity. Furthermore, the ecological integrity of small conservation areas can quickly be destroyed by edge effects caused by disturbances. The size structure of Zambian forest reserves is dominated by small areas of less than 110 km^2 (Figure 5.8). If 500 km^2 (Lebec and Goodland 1988) is taken as the minimum forest reserve size for biodiversity conservation, only 23 per cent of the Zambian forest reserves qualify as viable ecological units for biodiversity conservation. This observation supports the conclusion made by Botkin and Talbot (1992) that although the figure of protected areas in tropical countries is large, very few of these are self-sufficient ecological units. However, the majority of national parks in Zambia are large (Figure 5.8). These are therefore more appropriate as biodiversity conservation areas.

Nevertheless, size alone is not sufficient to ensure forest conservation. In national parks, the major threat to forest vegetation is destruction by large game animals. The major threats to conservation in forest reserves are deforestation, human encroachment and poor management. In the Zambian Copperbelt, about 8 per cent of the forest reserves had been clear-cut by 1984, while another 9 per cent had been converted to monoculture exotic plantation forests (Chidumayo 1989b). Although the recovery of indigenous forests following clear-cutting is good, especially under a controlled burning regime, little is known about recovery after plantation forestry.

The loss of forest reserves through illegal encroachment and legal de-reservation has also undermined the ecological integrity of forest reserves (Table 5.9). Although new additions may offset losses from de-reservation, the latter creates instability in forest reserves, as ecological systems for biodiversity conservation and monitoring, and should therefore be minimized or stopped altogether.

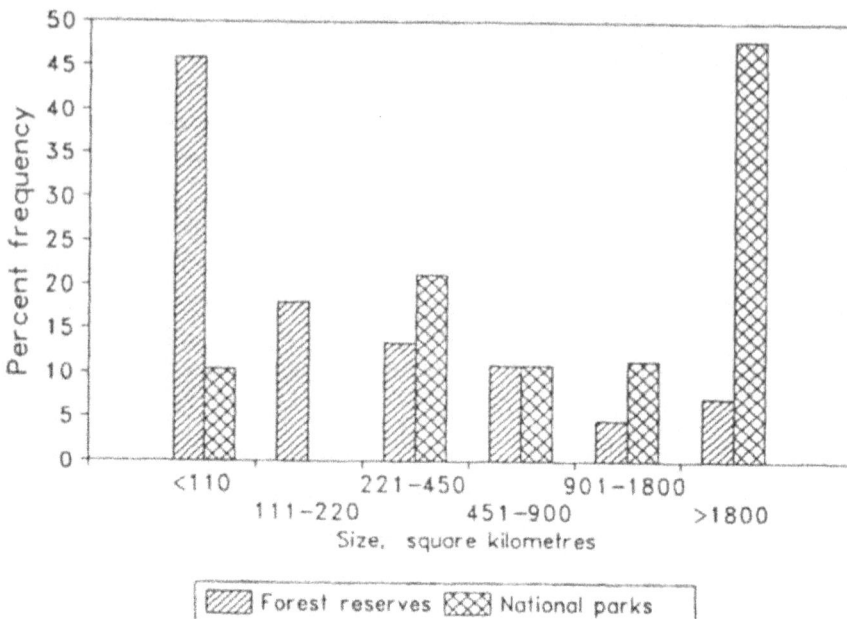

Figure 5.8 *Size distribution of forest reserves and national parks in Zambia*

Table 5.9 *Changes in the size of area under forest reserves in Zambia during 1978–87*

Year	Additions km²	Subtractions km²	Net change km²	Forest reserve area km²
1978	1 354	629	725	69 596
1979	1 647	125	1 521	71 117
1980	2 632	94	2 537	73 859
1981	548	95	454	74 307
1982	0	0	0	74 307
1983	339	6	332	74 639
1984	0	18	−18	74 622
1985	—	—	—	—
1986	1 519	1 365	154	73 783
1987	668	198	470	74 361

References

Adeyoju, S.K. 1991. *Forestry administrative institutions, manpower development and training, Zambia.* GRZ/FAO Consultant report (FO:DP/ZAM/88/C32: Field document no.2).
Allan, W. 1949. *Studies in African land usage in Northern Rhodesia.* The Rhodes-Livingstone papers No. 15. Institute for Social Research, Manchester University Press, Manchester.
Amphlett, M.B. 1986. *Soil erosion research project, Bvumbwe, Malawi.* Hydraulics Research, Wallingford (UK).
Anderson, J.M. and Ingram, J.S.I. (eds). 1989. *Tropical soil biology and fertility: A handbook of methods.* C.A.B. International, Wallingford (UK).
Araki, S. 1992. The role of miombo woodland ecosystem in chitemene shifting cultivation in northern Zambia. *Japan Infor MAB* 11: 8–15.
Astle, W.L. 1968–69. The vegetation and soils of Chishinga ranch, Luapula Province, Zambia. *Kirkia* 7: 73–102.
Bainbridge, W.R. and Edmonds, A.C.R. undated. *Forest Department management book for Kalomo and Livingstone.* Forest Department, Ndola.
Banda, M. 1988. *Epicormic and sucker shoots in miombo woodland management in Zambia.* M.Sc. dissertation in environmental forestry, University of North Wales, Bangor.
Bell, R.H.V. and Jachmann, H. 1984. Influence of fire on the use of *Brachystegia* woodland by elephants. *J. Afr. Ecol.* 22: 157–63.
Boaler, S.B. 1966. *The ecology of Pterocarpus angolensis in Tanzania.* Ministry of Overseas Development, London.
Boaler, S.B. and Sciwale, K.C. 1966. Ecology of a miombo site, Lupa North Forest Reserve, Tanzania. III: effects on the vegetation of local cultivation practices. *J. Ecol.* 54: 577–87.
Bonnor, G.M. 1987. Forest biomass inventory. In: Hall, D.O. and Overend, R.P. (eds), *Biomass: regenerable energy* (pp. 47–73), John Wiley and Sons, Chichester.
Botkin, B.D. and Talbot, L.M. 1992. Biological diversity and forests. In: Sharma, N.P. (ed.), *Managing the world's forests* (pp. 47–74), International Bank for Reconstruction and Development, Washington DC.
Boutette, M. and Karch, E.G. 1984. *Charcoal: small-scale production and use.* Deutsche Gesellschaft fur Technische Zusammenarbeit (GTZ), Eschborn. 60pp.

Bray, R.H. and Kurtz, L.I. 1945. Determination of the total organic and available forms of phosphorus in soils. *Soil Sci.* 59: 39–45.

Bremer, J.M. 1960. Determination of nitrogen in soils by the Kjeldahl method. *J. Agri. Sci.* 55: 1–23.

Brookman-Amissah, J., Hall, J.B., Swaine, M.D. and Attakorah, J.Y. 1980. A re-assessment of a fire protection experiment in north-eastern Ghana savanna. *J. Appl. Ecol.* 17: 85–99.

Bunyolo, A.M., Mbewe, D. and Malama, C. 1993. *Agriculture and soils.* Paper prepared for the National Environmental Action Plan for Zambia. Ministry of Environment and Natural Resources, Lusaka.

Campbell, B.M., Swift, M.J., Hatton, J. and Frost, P.G.H. 1988. Small-scale vegetation pattern and nutrient cycling in miombo woodland. In: Verhoeven, J.T.A., Heil, G.W. and Werger, M.J.A. (eds), *Vegetation structure in relation to carbon and nutrient economy* (pp. 69–85), SPB Academic Publishing, The Hague.

Campbell, B.M., Vermeulen, S.J. and Lynam, T. 1991. *Value of trees in the small-scale farming sector in Zimbabwe.* IDRC, Canada.

Celander, N. 1983. *Miombo woodlands in Africa – distribution, ecology and patterns of land use.* Swedish University of Agricultural Sciences (IRDC), Uppsala.

Chidumayo, N. (1979). Household woodfuel and environment in Zambia. Natural Resources Department, Lusaka.

Chidumayo, E.N. 1985. The ecology of urbanization and resource utilization: the case of Chipata district in eastern Zambia. Report prepared for Unesco MAB, Nairobi.

Chidumayo, E.N. 1986. *Bush fire and vegetation management in Zambia.* Natural Resources Department, Lusaka.

Chidumayo, E.N. 1987a. Species structure in Zambian miombo woodland. *J. Trop. Ecol.* 3: 109–118.

Chidumayo, E.N. 1987b. A shifting cultivation land use system under population pressure in Zambia. *Agrofor. Syst.* 5: 15–25.

Chidumayo, E.N. 1987c. Woodland structure, destruction and conservation in the Copperbelt area of Zambia. *Biol. Conserv.* 40: 89–100.

Chidumayo, E.N. 1988a. Regeneration of *Brachystegia* woodland canopy following felling for tsetse-fly control in Zambia. *Trop. Ecol.* 29: 24–32.

Chidumayo, E.N. 1988b. A re-assessment of effects of fire on miombo regeneration in the Zambian Copperbelt. *J. Trop. Ecol.* 4: 361–72.

Chidumayo, E.N. 1988c. Integration and role of planted trees in a bush-fallow cultivation system in central Zambia. *Agrofor.Syst.* 7: 63–76.

Chidumayo, E.N. 1989a. Early post-felling response of *Marquesia* woodland to burning in the Zambian Copperbelt. *J. Ecol.* 77: 430–8.

Chidumayo, E.N. 1989b. Land use, deforestation and reforestation in the Zambian Copperbelt. *Land Degradation and Rehabilitation* 1: 209–216.

Chidumayo, E.N. 1990. Above-ground woody biomass structure and productivity in a Zambezian woodland. *For. Ecol. Manage.* 36: 33–46.

Chidumayo, E.N. 1991a. Seedling development of the miombo woodland tree, *Julbernardia globiflora. J. Veg. Sci.* 2: 21–6.

Chidumayo, E.N. 1991b. Woody biomass structure and utilization for charcoal production in a Zambian miombo woodland. *Bioresource Technology* 37: 43–52.

Chidumayo, E.N. 1992a. Using natural fertilizer to sustain three traditional farming systems in miombowoodlands. Report prepared for the Biodiversity Support Program, World Wildlife Fund, Washington D.C.

Chidumayo, E.N. 1992b. Seedling ecology of two miombo woodland trees. *Vegetatio* 103: 51–8.

Chidumayo, E.N. 1992c. Effects of shoot mortality on the early development of *Afzelia quanzensis* seedlings. *J. Appl. Ecol.* 29: 14–20.

Chidumayo, E.N. 1992d. The utilization and status of dambos in southern Africa: a Zambian case study. In: Matiza, T. and Chabwela, H.N. (eds) Wetlands conservation conference for southern Africa (pp.105–108).

Chidumayo, E.N. 1993a. *Responses of miombo to harvesting: ecology and management*. Stockholm Environment Institute, Stockholm.

Chidumayo, E.N. 1993b. Zambian charcoal production – miombo woodland recovery. *Energy Policy* 21: 586–97.

Chidumayo, E.N. 1993c. Wood used in charcoal production in Zambia. Interim report for World Wildlife Fund (Biodiversity Support Programme), Washington D.C.

Chidumayo, E.N. 1993d. Silvicultural characteristics and management of miombo woodland. In: Piearce, G.D. and Gumbo, D.J. (eds.), *The ecology and management of indigenous forests in southern Africa* (pp. 124–133). Zimbabwe Forestry Commission and SAREC, Harare.

Chidumayo, E.N. 1994a. Inventory of wood used in charcoal production in Zambia. A report for the Biodiversity Support Program, World Wildlife Fund, Washington D.C.

Chidumayo, E.N. 1994b. Phenology and nutrition of miombo woodland trees in Zambia. *Trees* 9:67–72.

Chidumayo, E.N. 1994c. Effect of wood carbonization on soil and seedling productivity in miombo woodland. *For. Ecol. Manage* 70:353–357.

Chidumayo,E.N. and Chidumayo, S.B.M. 1984. The status and impact of woodfuel in urban Zambia. Department of Natural Resources, Lusaka.

Chidumayo, E.N. and Siwela, A. 1988. Utilization, abundance and conservation of indigenous fruit trees in Zambia. Paper presented at the ABN workshop on utilization and exploitation of indigenous and often neglected plants and fruits of eastern and southern Africa, Malawi, 21–27 August 1988.

Chilivumbo, A. and Kanyangwa, J. 1985. Women's participation in rural development programmes. University of Zambia (RDSB), Lusaka.

Christensen, J.M. 1985. Energy survey in Zambezi. Risö National Laboratory, Roskilde (Denmark).

Clark, D:J. 1975. Stone Age man at the Victoria Falls. In: Phillipson, D.W. (ed.), *Mosi-oa-Tunya: a handbook to the Victoria Falls region* (pp. 28–47), Longman, London.

Clauss, B. 1992. *Bees and beekeeping in the North-western Province of Zambia*. Mission Press, Ndola.

Cline-Cole, R.A., Main, H.A. and Nichol, J.E. 1990. On fuelwood consumption, population dynamics and deforestation in Africa. *World Development* 18: 513–27.

Cole, M.M. 1963. Vegetation and geomorphology in Northern Rhodesia: an aspect of the distribution of the savanna of central Africa. *Geog. J.* 129: 290–310.

Cunningham, A.B. 1993. *African medicinal plants: setting priorities at the interface between conservation and primary health care.* Unesco, Paris.

Dangerfield, J.M. 1993. Soil animals and soil fertility: a critical component of woodland productivity. In: Piearce, G.D. and Gumbo, D.J. (eds.) *The ecology and management of indigenous forests in southern Africa.* (pp. 209–215). Zimbabwe Forestry Commission and SAREC, Harare.

Day, P.R. 1965. Particle fractionation and particle size analysis. In: Clark, C.A. (ed.), *Methods of soil analysis* (pp. 545–67). American Society of Agronomy, Madison.

Department of Energy 1992. *Energy statistics bulletin: 1974–1990.* Ministry of Energy and Water Development, Lusaka.

Edmonds, A.C.R. 1964. *Forest Department management book for Lusaka and Feira districts.* Forest Department, Ndola.

Edmonds, A.C.R. 1976. *Vegetation map* (1:500,000) of Zambia. Surveyor General, Lusaka.

Egler, F.E. 1952. Vegetation science I. Initial floristic composition, a factor in old-field vegetation development. *Vegetatio* 4: 412–17.

Ellenbroek, G.A. 1987. *Ecology and productivity of an African wetland system: The Kafue flats, Zambia.* Dr W.Junk Publishers, Dordrecht.

Endean, F. 1968.*The productivity of miombo woodland in Zambia.* Forest (Department) Research Bulletin, Ndola.

Ernst, W. 1988. Seed and seedling ecology of *Brachystegia spiciformis*, a predominant tree component in miombo woodlands in south central Africa. *For. Ecol. Manage.* 25: 195–210.

Ernst, W. and Walker, B.H. 1973. Studies on the hydrature of trees in miombo woodland in south central Africa. *J. Ecol.* 61: 667–73.

Fanshawe, D.B. 1971. *The vegetation of Zambia.* Government Printer, Lusaka.

Forest Department/FAO 1986. *Wood consumption and resource survey of Zambia (Technical Note 2:The forest area).* Forest Department, Ndola.

Frost, P.G.H. 1985. The responses of savanna organisms to fire. In: Tothill, J.C. and Mott, J.J. (eds), *Ecology and management of the world's savannas* (pp. 232–7), Australian Academy of Sciences, Canberra.

Geary, T.F. 1972. Mukwa blight in central Africa. *Plant Disease Reporter* 56: 820–1.

Golley, F.B., McGinnis, J.T., Clements, R.G., Child, G.T. and Duever, M.J. 1975. *Mineral cycling in a tropical forest ecosystem.* University of Georgia Press, Athens. 248pp.

Greenland, D.J. and Kowal, J.M. 1960. Nutrient content of moist tropical forest. *Plant and Soil* 12: 154–74.

Guy, P.R. 1981. Changes in the biomass and productivity of woodlands in the Sengwa wildlife research area, Zimbabwe. *J. Appl. Ecol.* 18: 507–19.

Hayslett, H.T. 1967. *Statistics made simple.* W.H. Allen and Co. London.

Hesse, P.R. 1971. *A textbook of soil chemical analysis.* William Clowes and Sons, London.

Hibajene, S. 1994. *Assessment of Earth Kiln Charcoal Production Technology.* Energy, Environment and Development Series No. 39, Stockholm Environment Institute and Ministry of Energy, Lusaka

Högberg, P. 1992. Root symbioses of trees in African dry tropical forests. *J. Veg. Sci.* 3: 393–400.

Holden, S.T. 1988. *Farming systems and household economy in New Chambeshi, Old Chambeshi and Yunge villages near Kasama, Northern Province, Zambia: An agroforestry baseline study.* Zambia SPRP Studies Occasional Paper Series A (No. 9). Noragric, Agricultural University of Norway, ~As.

Hood, R.J. 1972. The development of a system of beef production for use in the Brachystegia woodlands of northern Zambia. Ph.D. thesis, University of Reading (Department of Agriculture), Reading.

Hudson, N. 1984. *Soil conservation.* English Language Book Society Publications, Oxford.

Hursh, C.R. 1960. *The dry woodlands of Nyasaland.* International Cooperation Administration (USA).

Jackson, R.M. and Raw, F. 1966. *Life in the soil.* Edward Arnold Publishers, London.

Kalumiana, O. 1994. *Charcoal supply stabilization project.* Department of Energy, Lusaka.

Kornas, J. 1978. Fire-resistance in pteridophytes of Zambia. *Fern Gazette* 11: 373–84.

Kotze, K. 1993. Study on the export potential of the Zambian handicraft industry. Report prepared for the European Development Fund, Lusaka.

Krebs, C.J. 1989. *Ecological methodology.* Harper and Row, New York.

Kwesiga, F. and Chisumpa, S.M. 1992. Multipurpose trees of the Eastern Province of Zambia: an ethnobotanical survey of their use in the farming systems. AFRENA Report 49. ICRAF, Nairobi.

Langworthy, H.W. 1971. Pre-colonial kingdoms and tribal migrations, AD 1500–1900. In: Davies, D.H. (ed.), *Zambia in maps* (pp. 32–3). University of London Press, London.

Lawton, R.M. 1978. A study of the dynamic ecology of Zambian vegetation. *J. Ecol.* 66: 175–98.

Lebec, G. and Goodland, R. 1988. *Wildlands: their protection and management in economic development.* The World Bank, Washington DC.

Lees, H.M.N. 1962. *Working plan for the forests supplying the Copperbelt, Western Province.* Government Printer, Lusaka.

Lewis, D. 1986. Disturbance effects on elephant feeding: evidence for compression in Luangwa Valley, Zambia. *J. Afr. Ecol.* 24: 227–41.

Lungu, O.I. and Chinene, V.R.N. 1993. Cropping and soil management systems and their effect on soil productivity in Zambia. Agricultural University of Norway (Ecology and Development Programme) As.

Macwani, M., Hibajene, S. and Mudenda, G. 1994. Domestic energy consumption and its impacts on the environment in Zambia. Report prepared for the African Development Bank (African Energy Programme), Abijan.

Malaisse, F. 1974. Phenology of the Zambezian woodland area with emphasis on the miombo ecosystem. In: Leith, H. (ed.), *Phenology and* seasonality modelling (*Ecological Studies* 8:269–286). Chapman and Hall, London.

Malaisse, F. 1978. The miombo ecosystem. In: *Tropical forest ecosystems* (pp. 589–606). Unesco/UNEP/FAO, Paris.

Malaisse, F. 1984. Structure of a Zambezian dry evergreen forest of the Lubumbashi surroundings (Zaire). *Bulletin de la Societé Royale de Botanique de Belgique* 177: 428–458.

Malaisse, F., Freson, R., Goffinet, G. and Malaisse-Mousset, M. 1975. Litter fall and litter breakdown in miombo. In: Golley, F.B. and Medina, E. (eds), *Ecological systems: trends in terrestrial and aquatic research (Ecological Studies* 11: 137–52). Springer Verlag, Berlin.

Marks, S.A. 1976. *Large mammals and a brave people: subsistence hunters in Zambia*. University of Washington Press, Seattle.

Marter, A. and Honeybone, D. 1976. The economic resources of rural households and the distribution of agricultural development. University of Zambia (RDSB), Lusaka.

Medwecka-Kornas, A. 1980. *Gardenia subaculus* Stapf and Hutch.: a pyrophytic suffrutex of the African savanna. *Acta Botanica Academiae Scientiarum Hungaricae* 26: 131–7.

Medwecka-Kornas, A. and Kornas, J. 1985. Fire-resistant sedges (*Cyparaceae*) in Zambia. *Flora* 176: 61–71.

Millington, A.C., Townsend, J.R.G., Saull, R.J., Kennedy, P. and Prince, S.D. 1986. SADCC fuelwood project: biomass assessment component. 2nd Interim Report, Munslow.

Moore, H.L. and Vaughan, M. 1994. *Cutting down trees*. Heinemann, Portsmouth, NH.

Moore, P.D. and Chapman, S.B. 1976. *Methods in plant ecology* (2nd edition). Blackwell Scientific Publications, Oxford.

Mumeka, A. 1986. Effect of deforestation and subsistence agriculture on runoff of the Kafue river headwaters, Zambia. *Hydrological Sciences J.* 31: 543–54.

Munyanziza, E. 1994. *Miombo trees and mycorrhizae: ecological strategies, a basis for afforestation*. Ph.D thesis, Wageningen Agricultural University, The Netherlands.

Musonda, F.B. 1986. Plant food in the diet of the prehistoric inhabitants of the Lunsenfwa drainage basin, Zambia, during the last 20 000 years. *Zambia Geographical Journal* 36: 17–27.

Mwamba, C.K. 1989. *An outlook on the role of indigenous fruit trees in agroforestry*. Paper presented at the first Zambian national agroforestry workshop, Lusaka.

Mwamba, C.K. in press. Effect of root-inhabiting fungi on root growth potential of *Uapaca kirkiana* (Muell. Arg.) seedlings. *Applied Soil Ecology* (in press).

Mwamba, C.K., Zgamba, Y. and Chongo, G. 1992. Effect of seedling source on post-planting growth of *Uapaca kirkiana* Muell. Arg. *S.Afr. For. J.* 161: 35–40.

Okali, D.U.U., Hall, J.B. and Lawson, G.W. 1973. Root distribution under a thicket clump on the Accra Plains, Ghana: its relevance to clump localization and water relations. *J. Ecol.* 61: 439–54.

Olsen, J. 1992. *The management of Faidherbia albida in agri-silvopastoral systems in southern Zambia.* M. Sc. dissertation in Environmental Forestry, University of North Wales, Bangor.

Palmer, J. and Synnott, T.J. 1992. The management of natural forests. In: Sharma, N.P. (ed.), *Managing the world's forests* (pp. 337–73). International Bank for Reconstruction and Development, Washington DC.

Pegler, D.N. and Piearce, G.D. 1980. The edible mushrooms of Zambia. *Kew Bull.* 35:475–91.

Pereira, H.C. 1975. *Land use and water resources.* Cambridge University Press, London.

Phillipson, D.W. 1971. Early man. In: Davies, D.H. (ed.), *Zambia in maps* (pp.28–31). University of London Press, London.

Piearce, G.P. 1993. Natural regeneration of indigenous trees: the key to their successful management. In: Piearce, G.D. and Gumbo, D.J. (eds.), *The ecology and management of indigenous forests in southern Africa* (pp. 109–123). Zimbabwe Forestry Commission and SAREC, Harare.

Ranta, J. and Makunka, J. 1986. *Charcoal from indigenous and exotic species in Zambia.* Forest (Department) Technical Note 29: 1–20.

Reeler, B., Campbell, B.C. and Price, L. 1991. Defoliation of *Brachystegia spiciformis* by a species-specific insect, *Melasoma quadralineata*, over two growing seasons. *J. Afr. Ecol.* 29: 271–4.

Richards, A.I. 1939. *Land, labour and diet in Northern Rhodesia: an economic study of the Bemba tribe.* International African Institute, Oxford University Press, London.

Robinson, D.A. 1978. *Soil erosion and soil conservation in Zambia: a geographical appraisal.* Zambia Geographical Association, Lusaka.

Rutherford, M.C. 1983. Growth rates, biomass and distribution of selected woody plant roots in *Burkea africana– Ochna pulchra* savanna. *Vegetatio* 55: 45–63.

Rutherford, M.C. and Panagos, M.D. 1982. Seasonal woody plant shoot growth in *Burkea africana-Ochna pulchra savanna. S. Afr. J. Bot.* 1: 104–116.

Savory, B.M. 1962. *Rooting habits of important miombo species.* Forest (Department) Research Bulletin 6: 1–120.

Scholes, R.J. 1990. The regrowth of *Colophospermum mopane* following clearing. *J. Grass. Soc. S. Afri.* 7: 147–51.

Schultz, J. 1974. *Explanatory study to the land use map of Zambia.* Ministry of Rural Development, Lusaka.

Scoones, I., Clark, J., Matose, F., Phiri, C., Hofstad, O., Makoni, I. and Mvududu, S. 1993. Future directions for forestry extension. In: Bradley, P.N. and McNamara, K. (eds), *Living with trees: policies for forestry management in Zimbabwe* (pp. 211–26). The World Bank, Washington D.C.

Sekeli, P.M. 1993. *Growing indigenous tree species in plantations.* Paper prepared for the miombo management handbook, Kitwe.

Serenje, W., Chidumayo, E.N., Chipuwa, J.H., Egneus, H. and Ellegård, A. 1994. *Environmental impact assessment of the charcoal production and utilization system in central Zambia.* Energy, Environment and Development Series No. 32, Stockholm Environment Institute, Stockholm.

Shackleton C.M. 1993. Fuelwood harvesting and sustainable utilization in a communal grazing land and protected area of the eastern Transvaal lowveld. *Biol. Conserv.* 63: 247–54.

Sharma, T.C. 1984. Some hydrological characteristics of the Zambian headwaters. *Zambia J. Sci. Techn.* 7: 12–21.

Sharma, T.C. 1985. Water resources research in Zambia review and perspectives. In: *Proceedings of the First National Fair on Science and Technology Research for Development* (pp. 30–9). National Council for Scientific Research, Lusaka.

Shea, R.W., Kauffman, J.B. and Shea, B.W. 1993. *Fuel biomass and combustion factors associated with fires in savanna ecosystems of South Africa and Zambia.* Report submitted to the USDA Forest Service, Intermountain Research Station, Missoula.

Soil Productivity Research Programme, 1987. *Annual research report, 1987.* Misamfu Regional Agricultural Research Station, Kasama.

Solbrig, O.T. (ed.), 1991. *Savanna modelling for global change.* Biology International Special Issue No. 24. IUBS, Paris.

Storrs, A.E.G. 1979. *Know your trees. Some of the common trees found in Zambia.* Forest Department, Ndola.

Storrs, A.E.G. 1982. *More about trees: a sequel to ' Know your trees'.* Forest Department, Ndola.

Stott, P. 1988. The forest as phoenix: towards a biogeography of fire in mainland south east Asia. *Geog. J.* 154: 337–50.

Strang, R.M. 1966. The spread and establishment of *Brachystegia spiciformis* (Beth.) and *Julbernardia globiflora* (Beth.) Troupin in the Rhodesian highveld. *Common. For. Rev.* 45: 253–6.

Strang, R.M. 1974. Some man-made changes in successional trends on the Rhodesian highveld. *J. Appl. Ecol.* 111: 249–63.

Stromgaard, P. 1984. The immediate effect of burning and ash-fertilization. *Plant and Soil* 80: 307–20.

Stromgaard, P. 1985a. A subsistence society under pressure: the Bemba of northern Zambia. *Africa* 55: 40–59.

Stromgaard, P. 1985b. Biomass estimation equations for miombo woodland, Zambia. *Agrofor. Syst.* 3: 3–13.

Stromgaard, P. 1985c. Biomass, growth and burning of woodland in a shifting cultivation area of south central Africa. *For. Ecol. Manage.* 12: 163–178.

Stromgaard, P. 1989. *Crop potential and adaptive strategies in Zambian agriculture.* University of Copenhagen (Institute of Geography), Copenhagen.

Stromgaard, P. 1990. Effects of mound-cultivation on concentration of nutrients in a Zambian miombo woodland soil. *Agri.Ecol. Env.* 32: 295–313.

Sweet, R.J. and Tacheba, G. 1985. Bush control with fire in semi-arid savanna in Botswana. In: Tothill, J.C. and Mott, J.J. (eds), *Ecology and*

management of the world's savannas (pp. 229–31). Australian Academy of Sciences, Canberra.
Swift, M.J., Heal, O.W. and Anderson, J.M. 1979 *Decomposition in terrestrial ecosystems* (Studies inEcology Vol. 5), Blackwell Scientific Publications, Oxford.
Tietema, T. 1993. Biomass determination of fuelwood trees and bushes of Botswana, southern Africa. *For. Ecol. Manage.* 60: 257–69.
Trapnell, C.G. 1953. *The soils, vegetation and agriculture of North-Eastern Rhodesia.* Government Printer, Lusaka.
Trapnell, C.G. 1959. Ecological results of woodland burning experiments in Northern Rhodesia. *J. Ecol.* 47: 129–168.
Trapnell, C.G. and Clothier, J.N. 1957. *The soils, vegetation and agricultural systems of North-Western Rhodesia.* Government Printer, Lusaka.
Trapnell, C.G., Friend, M.T., Chamberlain, G.T. and Birch, H.F. 1976. The effects of fire and termites on a Zambian woodland soil. *J. Ecol.* 64: 577–88.
van Gils, H. 1988. *Environmental profile: Western Province, Zambia.* ITC, Enschede.
Vernon, R. 1983. Field guide to important arable weeds of Zambia. Department of Agriculture, Mt Makulu Central Research Station.
Walkley, A. 1946. A critical examination of a rapid method for determining organic carbon in soils – effects of variations in digestion conditions and of inorganic soil constituents. *Soil Sci.* 63: 251–63.
Werger, M.J.A. and Coetzee, B.J. 1978. The Sudano-Zambezian region. In: Werger, M.J.A. (ed.), *Biogeography and ecology of southern Africa* (pp. 301–462), W. Junk Publisher, The Hague.
White, F. 1962. *The forest flora of Northern Rhodesia.* Oxford University Press, London.
White, F. 1976. The underground forests of Africa: a preliminary review. *Gardens' Bulletin* XXIX: 57–71.
White, F. 1983. *The vegetation of Africa.* Unesco, Paris.
World Bank, 1990. *Zambia urban household energy strategy.* ESMAP, Washington DC.
Zimba, S.C. 1991. *Sustainable fuelwood production at Kawambwa Tea Estate, Zambia.* M.Sc. thesis, University of Helsinki.

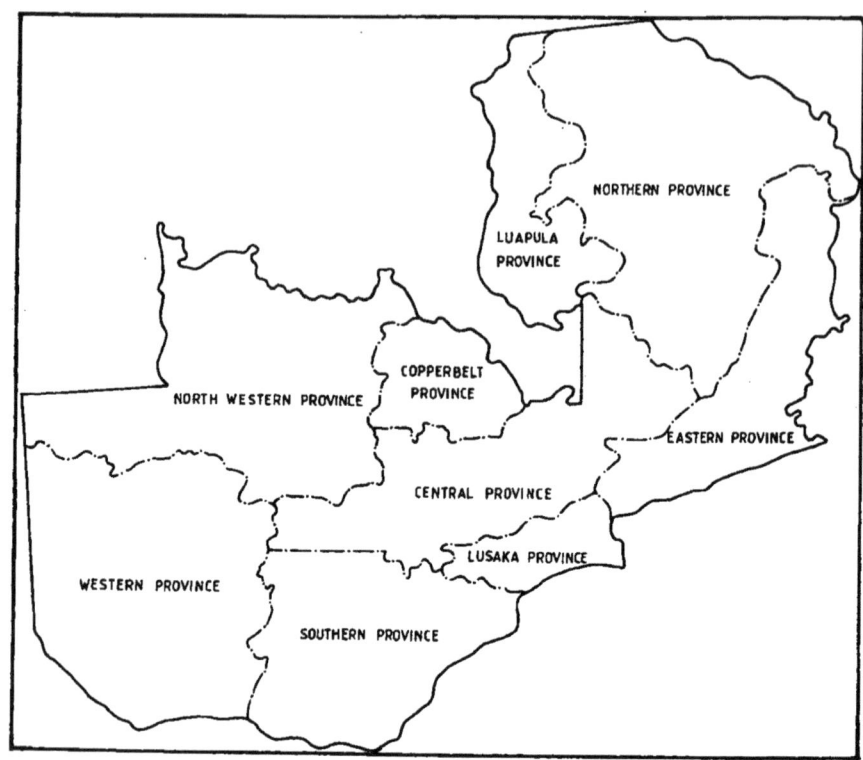

APPENDIX 1 *Map showing provincial divisions in Zambia*

APPENDIX 2 Biomass data for a 10 * 20 m dry miombo woodland plot clear-felled on 17 June 1991.

Tree stem	Species	GBH cm	Height m	1-m log	Mid girth cm	Fresh mass kg	Branch wood mass (kg) Live	Branch wood mass (kg) Dead	Total Sub-samples	Butt diam. mm	Minor branches Biomass (kg) Wood	Minor branches Biomass (kg) Leaf
11.1	Uapaca kirkiana	45	5.5	1	55	23.0	57.0	15.4	11	41	1.0	0.6
				2	46	17.8				40	2.8	0.4
				3	35	10.8						
11.2	Uapaca kirkiana	–		1	11	1.4			1	30	2.0	1.6
8	Parinari curatellifolia	36	2.9	1	40	12.2	7.2		8	35	2.0	0.6
			5.0	2	36	9.4				36	2.0	0.8
				3	29	6.4					2.2	
10	Burkea africana	28	5.7	1	33	9.0	2.4		14	45	4.6	2.0
				2	28	6.6				34	2.2	0.8
				3	20	3.4						
				4	22	3.8						
9	Faurea saligna	15	4.3	1	22	3.8			3	27	0.8	0.4
				2	16	2.2				37	2.8	1.8
4	Monotes sp.	6	2.3	1	49	20.0	11.8		1	32	1.0	0.4
1.1	Julbernardia globiflora	48	8.1	2	45	17.0			12	54	8.8	1.6
				3	45	17.0						
				4	44	16.6						
				5	45	17.0						
				6	28	8.2						
				7	24	4.6						
1.2	Julbernardia globiflora	–	4.4	1	48	19.4			1	34	3.2	1.2
				2	13	1.8						
7	Fauera saligna	10	3.7	1	15	2.0			5	24	1.2	1.0
				2	11	1.4						
5	Ochna sp.	66	7.8	1	67	33.4	91.0		13	63	7.0	0.8
				2	66	32.0				60	3.8	0.2
				3	44	14.2				23	0.2	0.2
				4	39	11.6						
				5	38	11.0						
				6	29	8.2						
6	Burkea africana	36	8.1	1	52	25.2	33.0		24	47	3.8	1.0
				2	60	20.0				28	2.0	0.6
				3	61	19.6						
2	Burkea africana	10	2.8	1	104	72.4	296.6		1	50	2.8	0.6
3	Uapaca nitida	90	9.3	2	94	67.0			90	39	2.2	0.6
				3	70	33.8				54	6.0	1.8
				4	64	32.0				30	1.6	0.8
				5	53	20.0						
				6	42	12.2						
				7	34	8.4						
				8	25	4.4						

APPENDIX 3   Equations for estimating leaf biomass.

		Species/group	Equation	r²
A. Young coppice shoots DBH ≥ 2.0 cm mass unit = g	1	Brachystegia boehmii	5.4BD-34	0.86
	2	Faurea speciosa	3.8BD-15	0.87
	3	Isoberlinia angolensis	5.3BD-28	0.84
	4	Julbernardia globiflora	4.7BD-28	0.91
	5	Uapaca kirkiana	3.3BD-7	0.92
	6	All species	4.6BD-23	0.82
B. Small stems DBH 3–11 cm mass unit = kg	7	Brachystegia boehmii	0.42DBH-1.16	0.83
	8	Isoberlinia angolensis	0.38DBH-1.47	0.83
	9	Julbernardia globiflora	0.41DBH-1.26	0.88
	10	Miombo genera	0.35DBH-0.79	0.93
	11	Uapaca kirkiana	0.32DBH-1.25	0.70
	12	Uapaca species	0.25DBH-0.69	0.69
	13	Other canopy species	0.26DBH-0.69	0.79
	14	Understorey species	0.16DBH-0.25	0.98
C. Large stems DBH > 11 cm mass unit = kg	15	Isoberlinia angolensis	0.50DBH-3.33	0.88
	16	Julbernardia globiflora	0.46DBH-3.28	0.70
	17	Miombo genera	0.35DBH-0.79	0.93
	18	Uapaca species	0.59DBH-5.27	0.96
	19	Other species	0.61DBH-5.86	0.96

APPENDIX 4   Equations for estimating twig wood biomass.

		Species/group	Equations	r²
A. Young coppice shoots DBH < 2.0 cm Mass unit =	1	Brachystegia boehmii	15BD-155	0.86
	2	Faurea speciosa	7BD-40	0.90
	3	Isoberlinia angolensis	3BD-14	0.88
	4	Julbernardia globiflora	9BD-60	0.90
	5	Uapaca kirkiana	6BD-43	0.84
	6	All species	11BD-9	0.74
B. Small stems DBH 3–11 cm Mass unit = kg	7	Brachystegia boehmii	1.49DBH-3.94	0.67
	8	Isoberlinia angolensis	1.60DBH-6.73	0.78
	9	Julbernardia globiflora	1.26DBH-3.89	0.81
	10	Miombo genera	1.51DBH-4.31	0.90
	11	Uapaca species	0.63DBH-0.97	0.96
	12	Other canopy species	0.62DBH-1.45	0.79
	13	Understorey species	0.24DBH-0.18	0.77
C. Large stems DBH > 11 cm Mass unit = kg	14	Miombo genera	1.92DBH-7.92	0.96
	15	Uapaca species	2.58DBH-16.29	0.91
	16	Other canopy species	1.22DBH-0.99	0.77
	17	Understorey species	1.23DBH-4.52	0.66

APPENDIX 5 *Equations for estimating cord wood biomass.*

	Species/group	Equation	r^2
A. Small stems DBH 3–11 cm	1 *Albizia* species	2.3DBH-5.7	0.82
	2 *Brachystegia boehmii*	0.5DBH-2.2H-7.2	–
	3 *Brachystegia spiciformis*	2.4DBH-7.8	0.88
	4 *Burkea africana*	2.7DBH-11.4	0.87
	5 *Dichrostachys cinerea*	0.5DBH+2.9H-9.4	–
	6 *Diplorhynchus condylocarpon*	0.9DBH-1.8	0.70
	7 *Faurea* species	1.3DBH-3.2	0.79
	8 *Isoberlinia angolensis*	0.8DBH+1.8H-7.6	–
	9 *Julbernardia globiflora*	2.3DBH-6.4	0.76
	10 *Julbernardia paniculata*	3.0DBH-9.0	0.70
	11 *Monotes* species	3.1DBH-9.2	0.81
	12 *Ochna* species	2.1DBH-7.1	0.88
	13 *Parinari curatellifolia*	1.9DBH-4.9	0.75
	14 *Phyllocosmus lemaireanus*	3.0DBH-11.5	0.71
	15 *Pericopsis angolensis*	2.0DBH-4.4	0.77
	16 *Protea* species	1.4DBH-3.9	0.89
	17 *Pseudolachnostylis maprouneifolia*	1.7DBH-4.1	0.85
	18 *Strychnos* species	1.4DBH-3.0	0.80
	19 *Swartzia madagascariensis*	2.5DBH-8.6	0.77
	20 *Syzygium guineense macrocarpum*	1.3DBH-3.3	0.73
	21 *Uapaca kirkiana*	1.0DBH+1.0H-5.3	–
	22 *Uapaca nitida*	1.7DBH+0.5H-7.1	–
	23 Other understorey species	1.4DBH-3.1	0.71
B. Large stems DBH > 11 cm	24 *Albizia* species	10DBH-88	0.93
	25 *Brachystegia boehmii*	10DBH-85	0.88
	26 *Brachystegia spiciformis*	44DBH-712	0.97
	27 *Isoberlinia angolensis*	15DBH-155	0.94
	28 *Julbernardia globiflora*	21DBH-229	0.87
	29 *Julbernardia paniculata*	27DBH-295	0.70
	30 *Parinari curatellifolia*	13DBH-129	0.94
	31 *Uapaca kirkiana*	14DBH-147	0.90
	32 *Uapaca nitida*	16DBH-164	0.84
	33 Other canopy species	16DBH-157	0.72
	34 Understorey species	6DBH-40	0.78

APPENDIX 6 *Biomass conversion factors for dry miombo woodland trees. SD is standard deviation and OD is oven-dry.*

Species	Mean	SD	Mean	SD	Mean	SD	Mean	SD
Albizia antunesiana	0.57	0.06	–	–	338	105	–	–
Brachystegia boehmii	0.57	0.04	557	61	371	89	0.45	0.13
B. spiciformis	0.60	0.03	589	61	292	38	0.54	0.06
B. utilis	0.62	0.03	–	–	–	–	0.39	0.19
Bridelia cathartica	0.59	0.02	502	26	–	–	–	–
Burkea africana	0.56	0.05	471	82	–	–	0.45	0.13
Dichrostachys cinerea	0.70	0.04	575	79	315	61	0.37	0.11
Diplorhynchus condylocarpon	0.55	0.02	403	68	286	87	–	–
Faurea spp.	0.58	0.04	517	83	–	–	0.36	0.11
Flacourtia indica	0.55	0.07	–	–	–	–	–	–
Hymenocardia acida	0.54	0.05	–	–	–	–	–	–
Isoberlinia angolensis	0.55	0.05	524	63	337	83	0.43	0.10
Julbernardia globiflora	0.61	0.05	626	64	349	118	0.50	0.09
Monotes spp.	0.60	0.03	579	95	308	48	0.48	0.07
Ochna spp.	0.54	0.03	447	28	292	54	–	–
Parinari curatellifolia	0.59	0.05	566	27	295	79	–	–
Pericopsis angolensis	0.55	0.04	547	81	336	103	0.30	0.11
Phyllocosmus lemaireanus	0.55	0.04	549	81	–	–	0.27	0.06
Protea gaudedi	0.49	0.05	452	61	348	41	–	–
Pseudolachnostylis maprouneifolia	0.54	0.03	531	64	348	64	–	–
Psorospermum febrifugum	0.56	0.05	–	–	–	–	–	–
Swartzia madagascariensis	0.63	0.02	–	–	258	52	–	–
Syzygium guineense macrocarpum	0.48	0.07	407	54	395	17	0.36	0.09
Uapaca kirkiana	0.49	0.04	469	85	297	75	0.30	0.07
Uapaca nitida	0.53	0.06	453	73	369	52	0.33	0.10

APPENDIX 7 Species structure of dry miombo occupying three habitats on the edge of Nyautai wet dambo in central Zambia.

Species	Number of stems			Above ground wood biomass (kg)		
	Creek zone	Wash zone	Scarp zone	Creek zone	Wash zone	Scarp zone
Bauhinia petersiana	1	0	0	19	0	0
Brachystegia bussei	1	0	0	385	0	0
Brachystegia longifolia	12	1	63	970	97	246
Brachystegia manga	9	0	0	1,807	0	0
Brachystegia spiciformis	0	10	0	0	155	0
Brachystegia utilis	0	0	38	0	0	609
Bridelia cathartica	2	0	0	11	0	0
Burkea africana	5	14	1	864	1,565	14
Cassia abbreviata	0	2	0	0	2	0
Commiphora sp.	0	2	0	0	14	0
Cussonia kirkii	0	0	7	0	0	143
Dalbergiellea nyassae	0	0	2	0	0	9
Dichrostachys cinerea	0	0	1	0	0	8
Diospyros botacana	0	0	1	0	0	0
Diplorhynchus condylocarpon	0	0	2	0	0	13
Faurea saligna	0	0	2	0	0	47
Flacourtia indica	1	1	0	0	0	0
Garcinia huillensis	0	0	1	0	0	0
Hexalobus monopetalus	0	6	0	0	30	0
Isoberlinia angolensis	15	0	0	5,071	0	0
Julbernardia globiflora	0	32	0	0	2,135	0
Lannea discolor	1	10	2	84	202	9
Monotes sp.	0	10	0	0	518	0
Ochna pulchra	0	9	0	0	43	0
Ochna scheinfurthiana	3	1	1	33	10	1
Parinari curatellifolia	4	42	13	524	961	258
Pericopsis angolensis	2	0	1	26	0	16
Phyllocosmus lemaireanus	5	0	4	162	0	49
Protea gaudedi	1	0	0	2	0	0
Pseudolachnostylis maprouneifolia	4	0	0	265	0	0
Psorospermum febrifegum	0	0	2	0	0	0
Rothmannia englerana	1	0	2	63	0	8
Securidaca longipedunculata	0	5	0	0	22	0
Strychnos innocua	0	11	0	0	136	0
Swartzia madagascariensis	0	6	0	0	185	0
Syzygium cordatum	19	7	40	2,270	267	1,404
Syzygium guineense macrocarpum	18	13	22	146	52	126
Terminalia brachystemma	0	4	0	0	282	0
Uapaca kirkiana	14	0	36	905	0	1,104
Uapaca nitida	6	10	63	944	709	2,214
Uapaca sansibarica	4	5	7	406	152	526
Unidentified species	3	1	2	314	0	110
Vagueriopsis lanciflora	0	2	0	0	7	0
Vitex madiensis	3	0	4	695	0	202
Total importance value	134	204	317	15,966	7,544	7,116
Total species (44)	23	23	24	23	23	24

APPENDIX 8 Species enumeration in contiguous (10 * 20 m) subplots in a dry miombo woodland stand for species-area curve analysis.

Species	Contiguous subplots									
	1	2	3	4	5	6	7	8	9	1–9
Julbernardia globiflora	4	7	10	5	21	12	14	26	10	109
Swartzia madagascariensis	7	1	0	0	0	2	2	0	1	13
Dichrostachys cinerea	4	4	3	1	0	1	5	11	2	31
Monotes sp.	1	3	2	2	0	1	1	0	0	10
Diplorhynchus condylocarpon	1	1	1	2	1	4	6	11	1	28
Uapaca nitida	1	0	0	0	0	0	0	0	0	1
Bridelia cathartica	1	0	2	2	0	2	1	0	1	9
Pseudolachnostylis maprouneifolia	2	0	3	0	2	3	4	2	0	16
Terminalia sp.		1	2	0	0	0	0	0	0	3
Pericopsis angolensis		3	0	3	0	0	1	0	0	7
Dalbergiellea nyassae		1	4	2	6	0	10	0	2	25
Brachystegia boehmii		4	2	4	6	4	0	0	2	22
Unidentified sp1		13	0	0	0	0	0	0	0	13
Cassia sp.		1	2	0	1	0	0	0	0	4
Unidentified sp2		1	0	0	0	0	0	0	0	1
Diospyros botacana		2	0	0	0	0	0	1	0	3
Flacourtia indica		1	0	0	2	2	1	1	1	8
Lannea sp.		1	0	0	0	2	1	1	1	6
Unidentified sp3			1	0	0	0	0	0	0	1
Brachystegia spiciformis			2	0	0	1	0	0	0	3
Ximenia sp.			1	0	0	2	2	4	0	9
Unidentified sp5				1	0	0	0	0	0	1
Afzelia quanzensis				1	0	0	0	0	0	1
Albizia antunesiana				1	1	0	0	1	0	3
Faurea speciosa					6	1	0	1	0	8
Vangueriopsis lanciflora					2	3	0	0	0	5
Unidentified sp6						1	0	0	0	1
Unidentified sp7							1	0	0	1
Hexalobus monopetalus							1	0	0	1
Psorospermum febrifegum									1	1
Burkea africana									1	1
Total stems	21	44	37	26	51	41	51	63	24	358
Species/plot	8	15	15	13	11	15	15	11	12	33
New species	8	10	5	3	2	1	2	0	2	33
Cumulative species	8	18	23	26	28	29	31	31	33	33
Cumulative area (ha)	.02	.04	.06	.08	1.0	1.2	1.4	1.6	1.8	1.8

APPENDIX 9 *Stem enumeration data for old growth wet miombo woodland plots (0.4 ha) before (in 1933:stems with gbh>15.2 cm) and after 11 years of fire management in 1944 (stems4.6 m) tall at Ndola, Zambia.*

Species	Before fire management in 1993			After 11 years of fire in 1944		
	2	1	3	2	1	3
Albizia antunesiana	2	0	3	2	0	2
Allophyllus africanus	0	2	5	0	0	0
Amblygonocarpus andongensis	1	0	1	0	0	0
Anisophyllea boehmii	9	6	12	7	6	9
Baphia bequaertii	14	5	5	13	5	9
Boscia corymbosa	1	0	1	0	0	1
Brachystegia longifolia	5	5	7	4	7	6
Brachystegia spiciformis	8	3	4	5	3	6
Bridelia cathartica	1	1	1	1	1	1
Burkea africana	2	1	1	0	0	0
Cassia abbreviata	0	1	3	0	0	0
Chrysophyllum bangweolense	1	0	0	1	0	0
Combretum molle + C.zehyeri	2	6	13	1	6	8
Dialiopsis africana	4	4	11	2	3	8
Dichrostachys cinerea	1	0	0	0	0	0
Diospyros batocana	9	7	11	9	4	7
Diospyros kirkii	0	2	1	0	0	0
Diplorhynchus condylocarpon	7	3	14	8	3	10
Ekebergia arborea	0	0	1	0	0	1
Erythrophleum africanum	3	3	6	3	3	7
Faurea saligna	0	0	1	0	0	0
Faurea speciosa	1	0	1	0	0	1
Ficus ingens	0	1	0	0	0	0
Flacourtia indica	0	1	3	0	0	0
Garcinia huillensis	0	1	0	0	1	0
Hexalobus monopetalus	4	3	8	0	0	3
Hymenocardia acida	1	1	1	0	2	1
Isoberlinia angolensis	19	17	8	13	16	6
Julbernardia paniculata	27	23	39	22	23	34
Lannea discolor	3	3	5	2	2	5
Lonchocarpus capassa	0	1	0	0	0	0
Maprounea africana	1	1	0	0	0	0
Markhamia obtusifolia	0	1	0	0	0	0
Monotes elegans	2	0	2	0	0	0
Ochna schweinfurthiana	4	5	3	3	4	1
Parinari curatellifolia	2	3	2	1	3	1
Parinari polyandra	1	0	1	1	0	1
Pericopsis angolensis	1	0	0	1	0	0
Phyllocosmus lemaireanus	2	3	2	1	1	1
Pseudolachnostylis maprouneifolia	2	6	5	1	3	7
Pterocarpus angolensis	1	0	3	1	0	3
Randia kuhniana	2	0	2	2	0	3
Strychnos cocculoides	2	1	3	0	0	1
Strychnos innocua	2	3	1	2	4	0
Strychnos pungens	1	1	1	1	1	1
Uapaca kirkiana	0	2	0	1	3	0
Uapaca nitida	3	0	0	3	0	0
Uapaca pilosa	1	1	0	0	0	0
Uapaca sansibarica	1	0	0	0	0	0
Unidentified species	4	2	1	3	4	9
Xylopia odoratissima	3	0	0	2	0	0
Ziziphus mucronata	1	1	0	0	0	0
Total stems	161	130	192	116	108	152
Species/plot	41	36	38	29	23	28

APPENDIX 10 Stem enumeration data in 0.4 ha wet miombo woodland plots before clear felling by stumping in 1933 (stems >20.3 cm gbh) and in regrowth under fire management in 1944 and 1982 (stems0 >.9 m tall) at Ndola, Zambia.

Species	Protected plot 4			Early burnt plot 2			Late burnt plot 3	
	1933	1944	1982	1933	1944	1982	1933	1944
Afzelia quanzensis	2	0	2	0	0	3	0	0
Albizia antunesiana + A.adianthifolia	1	12	12	1	6	8	2	0
Anisophyllea boehmii	13	16	1	7	26	3	5	7
Baphia bequaertii	36	107	17	16	50	17	19	9
Boscia corymbosa	0	4	0	0	5	0	0	0
Brachystegia longifolia	7	10	7	0	3	4	0	0
Brachystegia spiciformis	1	23	18	6	5	6	2	0
Bridelia cathartica	0	0	0	0	3	0	0	0
Bridelia duvigneaudii	0	15	0	0	6	0	0	0
Burkea africana	0	0	0	1	0	0	1	0
Byrsocarpus orientalis	0	120	0	1	20	0	0	2
Cathium crassum	0	24	0	0	7	0	0	0
Cassia abbreviata	1	0	0	1	0	0	0	0
Chrysophyllum bangweolense	0	0	0	1	0	0	0	0
Combretum molle + C.zeyheri	11	13	0	10	7	0	10	4
Dalbergia nitidula	1	0	0	0	0	0	0	0
Dialiopsis africana	0	26	0	5	32	0	10	3
Diospyros batocana	3	5	0	3	5	5	2	1
Diplorhynchus condylocarpon	6	11	3	4	8	5	6	16
Dombeya rotundifolia	0	0	0	0	0	0	0	1
Ekebergia arborea	0	4	0	0	0	0	0	1
Erythrophleum africanum	3	7	11	9	7	10	1	2
Faurea speciosa	0	11	0	0	3	0	0	2
Ficus ingens	1	0	1	0	0	2	0	0
Flacourtia indica	0	2	0	0	2	0	0	0
Garcinia huillensis	0	8	0	0	2	0	0	0
Hexalobus monopetalus	0	30	0	0	37	0	0	2
Hymenocardia acida	1	18	0	0	21	0	2	6
Indigofera rhynchocarpa	0	4	0	0	4	0	0	0
Isoberlinia angolensis	18	32	32	16	31	29	24	10
Julbernardia paniculata	56	27	20	91	36	53	69	4
Lannea discolor	2	12	3	2	6	5	2	2
Maprounea africana	0	13	0	0	16	0	0	2
Marquesia macroura	0	0	1	0	0	0	0	0
Memycylon flavovirens	0	36	0	0	6	0	0	0
Monotes spp.	1	0	0	2	0	3	1	0
Ochna schweinfurthiana	1	23	0	3	56	0	2	1
Ochna sp.	0	18	0	0	5	0	0	0
Other species	1	64	0	1	27	2	0	1
Parinari curatellifolia + P.polyandra	7	37	11	15	20	19	11	8
Pericopsis angolensis	2	8	6	0	1	1	1	0
Phyllocosmus lemaireanus	0	24	0	10	28	0	3	2
Protea homblei + P.trichophylla	0	2	0	0	2	0	0	0
Pseudolachnostylis maprouneifolia	0	12	8	1	24	14	1	1
Psorospermum febrifugum	0	76	0	0	147	0	0	4
Pterocarpus angolensis	0	10	0	1	6	0	1	27
Randia kuhniana	0	175	0	0	24	0	0	1
Rhothmannia anglerana	0	0	0	0	0	1	0	0
Rhus sp.	0	0	0	0	0	2	0	0
Strychnos cocculoides + S.spinosa	0	10	0	0	16	0	0	0
Strychnos innocua + S.pungens	2	32	2	13	30	5	5	6
Swartzia madagascariensis	1	2	0	0	8	1	0	1
Syzygium owariense + S.guineensis macrocarpum	1	2	0	0	8	1	1	2
Uapaca kirkiana	2	22	2	1	48	6	0	2
Uapaca nitida	0	8	0	0	1	0	1	4
Uapaca pilosa	0	2	0	0	3	0	0	2
Unidentified species	0	4	0	0	12	0	0	42
Vangueriopsis lanciflora	0	0	0	0	0	1	0	0
Vitex mombassae + V.madiensis	0	0	0	0	2	0	0	1
Xylopia adoratissima	0	24	0	0	3	0	0	3
Ziziphus mucronata	1	0	0	0	0	2	0	0
Total stems	82	1185	157	220	832	208	182	190
Species/plot	27	48	18	24	48	26	24	35

Appendix 11 *Broad guidelines for management of miombo regeneration*

Characteristics of miombo	Implications for management	
	Natural regeneration	Plantation regeneration
High species diversity	Management should not be species-specific but end-use related with an ecosystem approach	Monocrop planting may be unsuitable: guard against pests, disease and loss of symbiotic coexistence
Variable fruit and seed production	Leave seed trees when harvesting and allow adequate rotation cycles for seed production	Seed availability can be a problem
Seed viability is short and variable	Rely on seedlings and vegetative regrowth	Seed storage can be a problem
Seedlings develop a deep root before shoot development	Rely on coppice regrowth	Nursery root pruning can be damaging to seedlings
Seedlings are abundant in nature	Nurturing of seedlings may be important	Potential for transplanting wildlings
Slow growing from seed	Rely on coppice regrowth	Unsuitable for plantation but potential for vegetative propagation
Variability in growth rate and productivity	Yield and rotation generalizations may be misleading	Selection of species is important
High density in regrowth stands	Thinning may enhance growth of uncut stems	Spacing is necessary but may reduce productivity

www.ingramcontent.com/pod-product-compliance
Ingram Content Group UK Ltd.
Pitfield, Milton Keynes, MK11 3LW, UK
UKHW021842140426
5217IPUK00022B/1558